Phase-Transfer Catalysis

ACS SYMPOSIUM SERIES **326**

Phase-Transfer Catalysis
New Chemistry, Catalysts, and Applications

Charles M. Starks, EDITOR
Vista Chemical Company

Developed from a symposium sponsored by
the Division of Petroleum Chemistry, Inc.
at the 190th Meeting
of the American Chemical Society,
Chicago, Illinois,
September 8–13, 1985

American Chemical Society, Washington, DC 1987

Library of Congress Cataloging-in-Publication Data

Phase-transfer catalysis.
  (ACS symposium series; 326)

  "Developed from a symposium sponsored by the Division of Petroleum Chemistry, Inc. at the 190th Meeting of the American Chemical Society, Chicago, Illinois, September 8-13, 1985."

  Bibliography: p.
  Includes index.

  1. Phase-transfer catalysis—Congresses.
  2. Phase-transfer catalysts—Congresses.

  I. Starks, Charles M. II. American Chemical Society. Meeting (190th: 1985: Chicago, Ill.) III. American Chemical Society. Division of Petroleum Chemistry, Inc. IV. Series.

  QD505.P49   1987   541.3′95   86-25957
  ISBN 0-8412-1007-1

Copyright © 1987

American Chemical Society

All Rights Reserved. The appearance of the code at the bottom of the first page of each chapter in this volume indicates the copyright owner's consent that reprographic copies of the chapter may be made for personal or internal use or for the personal or internal use of specific clients. This consent is given on the condition, however, that the copier pay the stated per copy fee through the Copyright Clearance Center, Inc., 27 Congress Street, Salem, MA 01970, for copying beyond that permitted by Sections 107 or 108 of the U.S. Copyright Law. This consent does not extend to copying or transmission by any means—graphic or electronic—for any other purpose, such as for general distribution, for advertising or promotional purposes, for creating a new collective work, for resale, or for information storage and retrieval systems. The copying fee for each chapter is indicated in the code at the bottom of the first page of the chapter.

The citation of trade names and/or names of manufacturers in this publication is not to be construed as an endorsement or as approval by ACS of the commercial products or services referenced herein; nor should the mere reference herein to any drawing, specification, chemical process, or other data be regarded as a license or as a conveyance of any right or permission, to the holder, reader, or any other person or corporation, to manufacture, reproduce, use, or sell any patented invention or copyrighted work that may in any way be related thereto. Registered names, trademarks, etc., used in this publication, even without specific indication thereof, are not to be considered unprotected by law.

PRINTED IN THE UNITED STATES OF AMERICA

ACS Symposium Series

M. Joan Comstock, *Series Editor*

Advisory Board

Harvey W. Blanch
University of California—Berkeley

Alan Elzerman
Clemson University

John W. Finley
Nabisco Brands, Inc.

Marye Anne Fox
The University of Texas—Austin

Martin L. Gorbaty
Exxon Research and Engineering Co.

Roland F. Hirsch
U.S. Department of Energy

Rudolph J. Marcus
Consultant, Computers &
  Chemistry Research

Vincent D. McGinniss
Battelle Columbus Laboratories

Donald E. Moreland
USDA, Agricultural Research Service

W. H. Norton
J. T. Baker Chemical Company

James C. Randall
Exxon Chemical Company

W. D. Shults
Oak Ridge National Laboratory

Geoffrey K. Smith
Rohm & Haas Co.

Charles S. Tuesday
General Motors Research Laboratory

Douglas B. Walters
National Institute of
  Environmental Health

C. Grant Willson
IBM Research Department

Foreword

The ACS SYMPOSIUM SERIES was founded in 1974 to provide a medium for publishing symposia quickly in book form. The format of the Series parallels that of the continuing ADVANCES IN CHEMISTRY SERIES except that, in order to save time, the papers are not typeset but are reproduced as they are submitted by the authors in camera-ready form. Papers are reviewed under the supervision of the Editors with the assistance of the Series Advisory Board and are selected to maintain the integrity of the symposia; however, verbatim reproductions of previously published papers are not accepted. Both reviews and reports of research are acceptable, because symposia may embrace both types of presentation.

Contents

Preface... ix

1. Phase-Transfer Catalysis: An Overview............................... 1
   Charles M. Starks

2. Phase-Transfer Reactions Catalyzed by Metal Complexes................ 8
   Howard Alper

3. Mechanism of Phase-Transfer Catalysis: The Omega Phase............. 15
   Charles L. Liotta, Edward M. Burgess, Charles C. Ray,
   Elzie D. Black, and Barbara E. Fair

4. Cation-Binding Properties of Crown Ethers, Lariat Ethers, Bibracchial
   Lariat Ethers, and Poly(ethylene glycols) as Potential Phase-Transfer
   Catalysts... 24
   George W. Gokel, K. Arnold, T. Cleary, R. Friese, V. Gatto, D. Goli,
   C. Hanlon, M. Kim, S. Miller, M. Ouchi, I. Posey, A. Sandler,
   A. Viscariello, B. White, J. Wolfe, and H. Yoo

5. Stable Catalysts for Phase Transfer at Elevated Temperatures........... 38
   Daniel J. Brunelle

6. Reactivity and Application of Soluble and Polymer-Supported
   Phase-Transfer Catalysts... 54
   Fernando Montanari, Dario Landini, Angelamaria Maia, Silvio Quici,
   and Pier Lucio Anelli

7. Efficient Asymmetric Alkylations via Chiral Phase-Transfer Catalysis:
   Applications and Mechanism.. 67
   U.-H. Dolling, D. L. Hughes, A. Bhattacharya, K. M. Ryan, S. Karady,
   L. M. Weinstock, and E. J. J. Grabowski

8. The Phase-Transfer-Assisted Permanganate Oxidation of Alkenes and
   Alkynes... 82
   Donald G. Lee, Eric J. Lee, and Keith C. Brown

9. New Developments in Polymer Synthesis by Phase-Transfer Catalysis..... 96
   Virgil Percec

10. Mechanistic Aspects of Phase-Transfer Free Radical Polymerizations..... 116
    Jerald K. Rasmussen, Steven M. Heilmann, Larry R. Krepski, and
    Howell K. Smith II

11. Aromatic Substitution in Condensation Polymerization Catalyzed
    by Solid-Liquid Phase Transfer.................................... 128
    Raymond Kellman, Robert F. Williams, George Dimotsis,
    Diana J. Gerbi, and Janet C. Williams

12. Triphase Catalysis in Organometallic Anion Chemistry................ 143
    Robert A. Sawicki

13. The Scission of Polysulfide Cross-Links in Rubber Particles through Phase-Transfer Catalysis...................................155
    Paul P. Nicholas

14. Multisite Phase-Transfer Catalysts..................................169
    John P. Idoux and John T. Gupton

INDEXES

Author Index...187

Subject Index..187

Preface

SCIENTIFIC KNOWLEDGE AND INDUSTRIAL APPLICATION of the phase-transfer catalysis (PTC) technique have expanded greatly during the last 15 years. As its practical simplicity and economic advantages have become more recognized, this field appears to be growing even more. Growth of PTC is easily measured (1) by the number of entries indexed by *Chemical Abstracts*, (2) by the fact that *Chemical Abstracts* has decided to form a separate "C. A. Selects" on PTC, (3) by the increasing number of phase-transfer agents offered for sale by chemical supply companies, (4) by the worldwide increase in specialized publications on PTC in a variety of languages, and (5) by the less easily measured but clearly exponential growth in the use of PTC for industrial chemical processing. I estimate the volume of phase-transfer catalysts totaled about 40,000 lb of catalysts per year in 1980 but grew to more than 1 million lb of catalyst per year in 1985.

I am grateful to the authors who took the time and effort to prepare manuscripts, often on tight deadlines. I also wish to thank the many reviewers who helped to bring out points of particular interest, to clarify ambiguous areas, and to generally improve the quality of already excellent writing. I am especially grateful to Lisa Butler who single-handedly and cheerfully handled and prepared numerous letters, notices, copies, revisions, reviews, and manuscripts.

CHARLES M. STARKS
Vista Chemical Company
Ponca City, OK 74602

August 20, 1986

Chapter 1

Phase-Transfer Catalysis: An Overview

Charles M. Starks

Vista Chemical Company, Ponca City, OK 74602

Uniquely interesting, complex and useful activities and phenomena occur at interfaces: one need only to look at the interfaces between the land, the atmosphere, and the sea to find this truth. The same truth occurs in chemical interfaces, although sometimes it is the lack of activity that draws our attention. In many chemical situations where two species cannot collide and therefore cannot react because they are separated by an interface, the lack of activity has been overcome by use of the technique of **PHASE TRANSFER CATALYSIS (PTC)**, which not only allows reaction to occur, but often to occur in very selective ways.

An early and clear example of PTC(1) demonstrated that the lack of reactivity between a mixture of 1-chlorooctane and aqueous sodium cyanide (without organic solvent) could be overcome by the use of a <u>phase transfer agent</u>, whose function was to transfer cyanide ion in reactive form from its normal aqueous phase into the chlorooctane phase. Use of a small amount of phase transfer agent makes the system catalytic, since the phase transfer agent can repeatedly transfer active cyanide ions into the organic phase for reaction with 1-chlorooctane. This sequence of steps is represented by equation 1, where Q+ represents a quaternary salt containing sufficiently long alkyl groups or other organic structure as to make QCN predominantly soluble in the organic phase.

$$
\begin{array}{llll}
\text{organic phase} & C_8H_{17}Cl + Q^+CN^- \longrightarrow & C_8H_{17}CN + Q^+Cl^- \\
& \uparrow & \downarrow \\
\text{\textasciitilde} & & \downarrow \\
\text{aqueous phase} & Cl^- + Q^+CN^- \rightleftharpoons & CN^- + Q^+Cl^- \\
& \uparrow & \\
\end{array} \quad (1)
$$

Other classic examples illustrating the use of quaternary salts as phase transfer catalysts were published by Makosza(2), and by Brandstrom(3). Subsequent development of crown ethers(4-7) and cryptands(7-8) as phase transfer catalysts gave PTC an entirely new dimension since now the inorganic reagent, as sodium cyanide in the above equation, need no longer be dissolved in water but can be used as a dry solid. This development gave rise to liquid-solid PTC as a partner to liquid-liquid PTC.

Since 1971 phase transfer catalysis has emerged as a broadly useful tool(9-16), not only in organic chemistry, but also in inorganic chemistry(17), for new analytical applications(18), in electrochemistry(27a), photochemistry(27b), and especially in polymer chemistry.(21,27-31) The substantial number of publications, patents, reviews, and books (200 to 400 per year since 1980) concerned with PTC in both scientific and commercial applications attests to the high level of interest that this technique has generated.

Commercial usage of PTC techniques has increased markedly during the last five years not only in the number of applications (currently estimated to be fifty to seventy-five different uses(22)), but also in the volume of catalysts consumed (estimated to be about one million pounds per year(22)) and in the volume of products manufactured (estimated to be fifty to one hundred million pounds per year(22)) in the United States alone. Many indicators point to additional extensive commercial applications of the PTC technique all around the world, and these indicators suggest that future chemical manufacturing processes will more an more incorporate PTC because of its advantages of simplicity, reduced consumption of organic solvents and raw materials, mild reaction conditions, specificity of reactions catalyzed, and enhanced control over both reaction conditions, reaction rates, and yields. For some currently produced polymers PTC provides the only reasonable and practical commercial method of manufacture(22).

Enormous progress has been made in recent years in all aspects of phase transfer catalysis, and the symposium on which this volume is based(23) was organized to provide a sample of some of the advances in three areas: (1) theory and application of the method; (2) design of specific catalysts for increased efficiency and selectivity; and (3) use of PTC in polymer chemistry. The excellent chemists and their co-workers who have most generously contributed to this publication present a broad range of work and viewpoints which stimulate and delight those of us who have a strong interest in PTC.

Applications

The general concept of phase transfer catalysis applies to the transfer of any species from one phase to another (not just anions as illustrated above), provided a suitable catalyst can be chosen, and provided suitable phase compositions and reaction conditions are used. Most published work using PTC deals only with the transfer of anionic reactants using either quaternary ammonium or phosphonium salts, or with crown ethers in liquid-liquid or liquid-solid systems. Examples of the transfer and reaction of other chemical species have been reported(24) but clearly some of the most innovative work in this area has been done by Alper and his co-workers, as described in Chapter 2. He illustrates that gas-liquid-liquid transfers with complex catalyst systems provide methods for catalytic hydrogenations with gaseous hydrogen, catalytic oxidations with gaseous oxygen, and carbonylation of alkyl halides, olefins and acetylenes with carbon monoxide.

A portion of substantial theoretical and historical importance in PTC has centered around the question of 'what is the nature of the species transferred, and where in the reaction sequence do the slow steps take place.' These questions prompted much debate during the early years of PTC when development and understanding of the reaction sequence and mechanism of PTC was emerging. It generally came to be recognized that small changes in the complex sequence of steps of even the simplest phase transfer catalyzed system can result in differing kinetics, differences in optimal catalyst structure, and different optimal reaction conditions, even for similar reactions. Of particular interest in this volume is Chapter 3 by Liotta and co-workers, who have found that even the amount of water present in liquid-solid PTC may substantially affect the site where final reaction occurs. These observations have led Liotta to postulate a new phase in which reaction may occur. Montanari and co-workers, who have been prolific contributors in PTC, have also provided (Chapter 6) significant insight into the effect of water on PTC reactions.

In a series of highly useful applications phase transfer catalysis has given the chemist the ability to conduct reactions between organic compounds and strong inorganic oxidants such as permanganate, dichromate, hypochlorite, and hydrogen peroxide(25). Use of these oxidants previously has been limited and experimentally inconvenient because of the narrow range of stable organic solvents which one could use to bring the oxidants and substrate into contact. The power of PTC for permanganate oxidations and the use of PTC to study the mechanism of these oxidations is demonstrated by Lee in Chapter 8.

Catalyst Improvements

Although quaternary ammonium salts, phosphonium salts, crown ethers, and cryptands are generally excellent catalysts for most PTC applications, there are many potential PTC applications where these agents have deficiencies. For example, ordinary tetraalkyl quaternary salts decompose at high temperatures (greater than 120-149°C), or at low temperatures under highly alkaline conditions or in the presence of highly nucleophilic anions such as phenoxide. Crown ethers and cryptands are stable under these conditions but are much more expensive (100 to 1000 fold cost) than quaternary salts, and for practical commercial use they must be completely recovered for re-use. Open-chain polyethers, mainly polyethylene glycols, catalyze some kinds of two-phase reactions, are stable and are quite low in cost, but their limited utility substantially restricts their application to a narrow range of reactions.

The possibility of solving the catalyst recovery problem by attaching active catalyst centers to insoluble polymeric substrates was recognized early(26), as was the possible use of chiral PTC catalysts to introduce chirality in products(1). Much work in both these areas has been partially successful(27). However, the results have not been completely satisfactory in that resin bound catalysts have shown much lower catalytic activity than soluble catalysts and they frequently lose their activity with repeated use. Chiral catalysts generally give products low chiral selectivity, with enantiomeric excesses usually less than fifty percent. A further problem in PTC has been that no really superior catalysts have been introduced for use with divalent or trivalent anions, or with the difficult transferable hydroxide anion.

In face of the above discouraging results, recent innovative catalyst work has led to highly effective solutions for some otherwise very difficult and expensive problems. For example, Dolling and co-workers (Chapter 7) have shown that by careful choice of PTC catalyst and use of optimal reaction conditions one can obtain high chiral selectivity (greater than 90% enantiomeric excess) and have applied this chemistry to a commercial process for production of the diuretic drug candidate Indacrinone.

Brunelle, in Chapter 5, has provided a solution to the problem of quaternary ammonium catalysts being unstable at elevated temperatures in the presence of highly nucleophilic anions. He found that catalysts based on p-dialkylaminopyridinium salts are approximately one hundred times more stable than simple tetraalkylammonium salts and are useful even up to temperatures of 180°C. Especially valuable is the fact that under these conditions a variety of nucleophilic displacement reactions on aryl halides occurs, making possible the economical commercial synthesis of otherwise difficulty available poly aryl ethers and sulfides.

Brunelle, Chapter 5, also demonstrated that bis-quaternary salts with appropriate spacing between the quaternary nitrogens are dramatically better than mono-quaternary salts as catalysts for transfer of divalent anions, such as the di-anion of bisphenol A. Thus the ion pair formed from the di-anion and a bis-quat appears to be more easily formed and transferred than the species formed from the di-anion and two mono-quaternary cations.

Idoux and co-workers, Chapter 14, have also prepared high-activity multi-site phase transfer catalysts bound to insoluble resins. Although not yet experimentally demonstrated, this type of catalyst may also be useful for transferring multivalent anions such as carbonate, sulfite, sulfide, sulfate or phosphate, since the compounds used have two phosphonium cationic centers in close proximity to each other. However, these workers present work which shows that a multi-site catalyst can exhibit greater selectivity in displacement on organic reagents containing two displaceable groups.

Montanari and co-workers, Chapter 6, have developed special cyclic ethers which, when properly bound to cross-linked resins, exhibit a high degree of catalytic effectiveness, generally comparable to that of soluble quaternary ammonium and phosphanium salts. This great increase in activity for resin-bound catalysts represents a breakthrough development, and although these catalysts would be highly expensive, their ability to maintain high activity after repeated use and for extended times in continuous reactors would mitigate their initial cost and avoid the problem of catalyst removal from products.

Taking an entirely different tack on supported phase transfer catalysts, Sawicki, Chapter 12, initially used polyethers chemically bound to silica. But, he also demonstrated that solid silica or alumina alone may function as liquid-solid phase transfer catalysts, probably through mechanisms entirely different than the classical PTC sequence.

Exploration by Gokel and co-workers, Chapter 4, of a diverse set of polyether structures and their ability to bind with cations has provided compounds which have many potential applications for use in PTC.

Phase Transfer Catalysis in Polymer Chemistry

Aside from the use of polymers as supports for phase transfer catalyst centers, much excellent work has been reported on the use of PTC in polymer chemistry for polymerization methods(28), for the chemical modification of already formed polymers(29), for the modification of polymer surfaces without change of the bulk polymer(30), and for the preparation and purification of monomers(31).

Rasmussen and co-workers, Chapter 10, have shown that many free-radical polymerizations can be conducted in two-phase systems using potassium persulfate and either crown ethers or quaternary ammonium salts as initiators. When transferred to the organic phase persulfate performs far more efficiently as an initiator than conventional materials such as azobisisobutyronitrile or benzoyl peroxide. In vinyl polymerizations using PTC-persulfate initiation one can exercise precise control over reaction rates, even at low temperatures. Mechanistic aspects of these complicated systems have been worked out for this highly useful and economical method of initiation of free-radical polymerizations.

Production of polymers through poly-substitution or poly-condensation reactions would be expected to be a natural extension of simple PTC chemistry. To a large extent this is true, but as Percec has shown, Chapter 9, the ability to use two-phase systems for these reactions has enormously extended the chemist's ability to control the structure of the polymers produced. Kellman and co-workers (Chapter 11) have also extensively studied poly-substitution displacements on perfluorobenzene substrate to produce unique polymers.

The simple reaction of polymers with inorganic reagents has historically been a difficult chemical problem and generally such transformations have been far too expensive to practice on a commercial scale. The use of PTC makes this kind of problem vastly simpler, as innovatively demonstrated in an adaptation by Nicholas (Chapter 13), to find low-cost chemistry for converting scrap rubber into a material more nearly resembling the structure of new rubber.

PTC Development

About ten years ago a knowledgeable organic chemist offered the opinion that "almost all the things that can be done via phase transfer catalysis has already been done." He was wrong, of course, as one can now look back and see that the great bulk of PTC chemistry now known came after his comment was made. While it may be true that many of the obvious and direct applications of PTC, especially for anion transfer, have been identified, it seems most likely to this author that a vast amount of new applications and more complex catalyst systems based on PTC await discovery and exploitation.

References

1. C. M. Starks, J. Am. Chem. Soc., 93, 195 (1971).
2. M. Makosza, "Reactions of Carbanions and Halogenocarbenes in Two-Phase Systems," Russian Chem. Revs. 46, 1151 (1977); "Two-Phase Reactions in Organic Chemistry" in Survey of Progress in Chemistry, Vol. IX, (1979) Academic Press, New York, and references contained therein.
3. A Brandstrom, "Principles of Phase-Transfer Catalysis by Quaternary Ammonium Salts," Adv. Phys. Org. Chem. 15, 267 (1977).
4. A. W. Herriott and D. Picker, J. Am. Chem. Soc. 97, 2345 (1975).
5. C. L. Liotta and H. P. Harris, J. Am. Chem. Soc. 95, 2250 (1974).
6. G. W. Gokel, D. J. Cram, C. L. Liotta, H. P. Harris, and F. L. Cook, J. Org. Chem. 39, 2445 (1974).
7. F. Montanari, Chim. Ind. (Milan) 57, 17 (1975).
8. R. Fornasier and F. Montanari, Tetrahedron Lett. 1381 (1976).
9. A Brandstrom, "Preparative Ion Pair Extraction," Apotekarsocieteeten/Hassle Lakemedel, Sweden, 1974.
10. W. P. Weber and G. W. Gokel, "Phase Transfer Catalysis in Organic Synthesis," Springer Verlag, New York 1977.
11. R. A. B. Bannard, "Phase Transfer Catalysis and Some of its Applications to Organic Chemistry," U.S. Dept. of Commerce NITS AD-A-030 503, July, 1976.
12. E. A. Dehmlow and S. S. Dehmlow, "Phase Transfer for Catalysis," Chem. Verlag 1983
13. C. M. Starks and C. L. Liotta, "Phase Transfer Catalysis, Principles and Techniques," Academic Press, New York, 1978.
14. J. M. McIntosh, "Phase Transfer Catalysis Using Quaternary Onium Salts," J. Chem. Educ. 55, 235 (1978).
15. G. W. Gokel and W. P. Weber, "Phase Transfer Catalysis," J. Chem. Educ. 55, 350 (1978).
16. W. E. Keller, "Compedium in Phase Transfer Reactions and Related Synthetic Methods," Fluka, Switzerland (1979).
    H. H. Freedman, "Industrial Applications of Phase Transfer Catalysis, Past, Present, and Future," in press.
17. R. M. Izatt and J. J. Christensen, "Synthetic Multidentate Macrocyclic Compounds," Academic Press, New York, 1978.
18. N. A. Gibson and J. W. Hosking, Aust. J. Chem. 18, 123 (1965).
19. For electrochemical applications of PTC see the following and references contained therein: E. Laurent, R. Rauniyar, and M. Tomalla, J. Appl. Electrochem. 14, 741 (1984): 15, 121 (1985). S. R. Ellis, D. Pletcher, W. M. Brooks, and K. P. Healy, J. Appl. Electrochem., 13, 735 (1983). Asahi Chemical Industry Co., Ltd., Japanese Patent Kokai 58/207382 (1982).
20. For photochemical applications of PTC see the following references and references contained therein: Z. Goren and I. Willner, J. Am. Chem. Soc. 105, 7764 (1983). T. Kitamura, S. Kobayashi, and H. Taniguchi, J. Org. Chem. 49, 4755 (1984).
21. F. L. Cook and R. W. Brooker, "Polymer Syntheses Employing Phase Transfer Catalysis," Polym. Prepr., Am. Chem. Cos., Div. Polym. Chem., 23, 149 (1982).

22. Estimates of commercial use were compiled by the author from a collection of non-published sources.
23. Symposium on Advances in Phase Transfer Catalysis, American Chemical Society Meeting, September 9, 1985, Chicago, sponsored by the Petroleum Chemistry Division.
24. A. Suzuki, T. Nakata, and W. Tanaka, Japanese Patent 70/10,126; C.A. 73, 44885 (1970). L. O. Esayan and Sh. O. Badanyan, Arm. Khim. Zh. 28, 75 (1975); C. A. 83, 9062 (1975). P. S. Hallman, B. R. McGarvey, and G. Wilkinson, J. Chem. Soc. A, 3143 (1968). H. M. van Dort and H. J. Geurse, Rec. Trav. Chim. Pays-Bas 86, 520 (1967).
25. For application of PTC with strong inorganic oxidants, see the following references and references contained therein: O. Bortolini, F. Di Furia, G. Modena, and R. Seraglia, J. Org. Chem. 50, 2688 (1985). H. E. Fonouni, S. Krishnan, D. G. Kuhn, and G. A. Hamilton, J. Am. Chem. Soc., 105, 7672 (1983). H. Alper, Adv. Organomet. Chem., 19, 183 (1981).
26. S. L. Regan, J. Am. Chem. Soc., 97, 5956 (1975).
27a. For reviews on the use of polymer-supported phase transfer catalysts, see: W. T. Ford, Adv. Polym. Sci. 55, 49 (1984); Polym. Sci & Tech., 24, 201 (1984). D. C. Sherrington, Macromol. Chem. (London) 3, 303 (1984).
27b. For use of chiral phase transfer catalysts see the following references and references contained therein: J. W. Verbicky Jr., and E. A. O'Neil, J. Org. Chem., 50, 1786 (1985); E. Chiellini, R. Solaro, and S. D'Antone, Polym. Sci. Technol. (Plenum), 24 (Crown Ethers Phase Transfer Catal. Polym. Sci.) 227 (1984).
28. For use of phase transfer catalysts in polymerization reactions see the following references and references contained therein: Y. Imai and M. Ueda, Polym. Sci. Technol. (Plenum), 24 (Crown Ethers Phase Transfer Catal. Polym. Sci.) 121 (1984); R. Bacskai, ibid 183; F. L. Cook and R. W. Brooker, Polym. Prepr. (Am. Chem. Soc., Div. Polym. Chem), 23, 149 (1982); C. E. Carraher, Jr., and M. D. Naas, ibid 158; A. Jayakrishnan and D. O. Shah, J. Polym. Sci., Polym. Chem. E., 21, 3201 (1983); J. Appl. Polym. Sci., 29, 2937 (1984).
29. For phase transfer catalyzed chemical reactions of polymers see the following references and references contained therein: J. M. J. Frechet, Polym. Sci. Technol (Plenum), 24 (Crown Ethers Phase Transfer Catal. Polym. Sci.), 1 (1984); G. Martinez, P. Terroba, C. Mijangos, and J. Millan, Rev. Plast. Mod., 49, 63 (1985); F. F. He and H. Kise, Makromol. Chem., 186, 1395 (1985); W.H. Daly, J. D. Caldwell, V. P. Kien, and R. Tang, Polym. Prepr. (Am. Chem Soc., Div. Polym. Chem.) 23, 145 (1982); D. C. Sherrington, Macormol. Chem. (London), 3, 303 (1984).
30. For use of phase transfer catalysis in modification of the surface of polymers see: A. J. Dias and T. J. McCarthy, Polym. Mater. Sci. Eng., 49, 574 (1983); Macromolecules, 17, 2529 (1984); H. Kise and H. Ogata, J. Polym. Sci., Polym. Chem. Ed., 21, 3443 (1983).
31. For an example of the use of phase transfer catalysis in monomer purification see: J. T. Fenton, U.S. Patent 4,423,238 (1983).

RECEIVED August 12, 1986

Chapter 2

Phase-Transfer Reactions Catalyzed by Metal Complexes

Howard Alper

Department of Chemistry, University of Ottawa, Ottawa, Ontario, Canada K1N 9B4

Recent studies indicate that phase transfer catalysis is useful for effecting a variety of interesting metal catalyzed reactions. Developments in the author's laboratory, in three areas, will be considered: reduction, oxidation, and carbonylation reactions.

Reduction

Nitro Compounds

Aliphatic and aromatic amines can be obtained in excellent yields by the ruthenium carbonyl catalyzed reduction of nitro compounds using carbon monoxide, aqueous base, benzene or toluene as the organic phase, and benzyltriethylammonium chloride as the phase transfer agent. This reaction occurs at room temperature and 1 atmosphere pressure(1). The ruthenium(II) complex, $RuCl_2(PPh_3)_3$, can also be used as the catalyst for this transformation, although product yields are highest when synthesis gas is used instead of carbon monoxide(2).
    Although neither cobalt carbonyl nor chloro(1,5-hexadiene)rhodium(I) dimer (or other rhodium(I) complexes) were effective for the reduction of nitro compounds, the use of both catalysts and phase transfer conditions resulted in the formation of amines in good yields(3). Subsequent studies demonstrated

$$ArNO_2 \xrightarrow[C_{12}H_{25}N(CH_3)_3{}^+Cl^-, \text{ r.t., 1 atm.}]{CO, 5N\ NaOH,\ C_6H_6, Co_2(CO)_8, [1,5-HDRhCl]_2} ArNH_2 \quad (1)$$

that this apparent bimetallic phase transfer process is a consequence of a novel coincidence effect. That is, the nitro reduction is catalyzed by the rhodium(I) complex alone under biphasic conditions but the presence of a phase transfer agent results in inhibition of the reaction. Reactivation of the system is achieved by addition of the second metal catalyst, cobalt carbonyl(4).

2. ALPER *Metal Complex Catalysis*

Mercaptans

Activated mercaptans undergo desulfurization to hydrocarbons using cobalt carbonyl or triiron dodecacarbonyl as the metal complex, and basic phase transfer conditions(5). Acidic phase transfer catalysis has been little investigated, the first example in organometallic chemistry being reported in 1983 (reduction of diarylethylenes)(6). When acidic phase transfer conditions (sodium 4-dodecylcenzenesulfonate as the phase transfer catalyst) were used for the desulfurization of mercaptans [$Fe_3(CO)_{12}$ as the metal complex],

$$R_1R_2R_3CSH \xrightarrow[\substack{C_6H_6,\ 60°,\ 1\ atm.\\ p-C_{12}H_{25}C_6H_4SO_3Na}]{Fe_3(CO)_{12},\ 48-50\%\ HBF_4} R_1R_2R_3CH \quad (2)$$

hydrocarbons were obtained in modest yields together with sulfides and disulfides. Good to excellent yields of hydrocarbons were realized when the reaction was effected in the absence of the phase transfer agent (i.e., as a biphasic process). Some results are listed in Table I(7).

Table I. Biphasic and Acidic Phase Transfer Catalyzed Reactions of of Mercaptans with $Fe_3(CO)_{12}$ and 48-50% HBF_4

Mercaptan	Products	Yield(%)	
		Phase Transfer	Biphasic
$p-CH_3OC_6H_4CH_2SH$	$p-CH_3OC_6H_4CH_3$	10	72
$p-ClC_6H_4CH_2SH$	$p-ClC_6H_4CH_3$	6	44
	$(p-ClC_6H_4CH_2)_2S$	11	19
	$(p-ClC_6H_4CH_2S)_2$	40	6
$2,4-Cl_2C_6H_3CH_2SH$	$2,4-Cl_2C_6H_3CH_3$	73	74
	$(2,4-Cl_2C_6H_3CH_2)_2$	5	10
	$(2,4-Cl_2C_6H_3CH_2)_2S$	2	2
	$2,4-Cl_2C_6H_3CHO$	3	3
$(p-CH_3C_6H_4)_2CHSH$	$(p-CH_3C_6H_4)_2CH_2$	-[a]	94
Ph_2CHSH	Ph_2CH_2	-[a]	84
	$(Ph_2CH)_2S$		8

[a] Not done

Aromatic Hydrocarbons and Heterocyclic Compounds

The dimer of chloro(1,5-hexadiene)rhodium is an excellent catalyst for the room temperature hydrogenation of aromatic hydrocarbons at atmospheric pressure. The reaction is selective for the arene ring in the presence of ester, amide, ether and ketone functionalities (except acetophenone). The most useful phase transfer agents are tetrabutylammonium hydrogen sulfate and cetyltrimethylammonium bromide. The aqueous phase is a buffer of pH 7.6 (the constituents of the buffer are not critical). In all but one case the reaction is stereospecific giving cis products

$$\text{o-CH}_3\text{O-C}_6\text{H}_4\text{-COOCH}_3 \xrightarrow[\text{C}_6\text{H}_{14}, \text{ buffer} \atop \text{r.t., 1 atm.}]{\text{H}_2, [1,5\text{HDRhCl}]_2} \text{cis-2-methoxy-cyclohexyl methylcarboxylate} \quad (3)$$

The hydrogenation reaction is also applicable to a range of heterocyclic systems including furans, pyridines and quinolines(8).

Oxidation

An industrially important oxidation reaction, the palladium catalyzed conversion of ethylene to acetaldehyde, is known as the Wacker process. While this transformation works well for ethylene and also propylene, it has not been applied commercially to the production of ketones from longer chain olefins including butenes. It seemed conceivable to us that the use of phase transfer techniques would permit the oxidation to occur under gentle conditions. Indeed, terminal olefins, including 1-butene, are oxidized to ketones by oxygen, palladium chloride as the metal catalyst, cupric chloride as re-oxidant, benzene, water and a quaternary ammonium salt as the phase transfer agent (80°C, 1 atm). This reaction proceeds only when the phase transfer catalyst contains at least one long-chain alkyl group (e.g., cetyltrimethylammonium bromide)(9). Complexes of rhodium and ruthenium also catalyze the same conversion, albeit in lower yields (10). While these reactions have the advantage of selectivity to terminal olefins, the failure to oxidize 2-butene (and other internal olefins) is a drawback.

Polyethylene glycols (PEG) have been employed as phase transfer agents (and as solvents) in a number of reactions(11). Application of PEG-400 to the Wacker reaction results in the oxidation of both terminal and internal olefins (e.g., isomeric butenes to butanone)(12).

$$C_2H_5CH=CH_2 / CH_3CH=CHCH_3 \xrightarrow[\text{PEG-400}, H_2O, 65°]{O_2, PdCl_2, CuCl_2 \cdot 2H_2O} C_2H_5COCH_3 \quad (4)$$

2. ALPER   Metal Complex Catalysis

An important class of inclusion compounds are cyclodextrins, which are cyclic oligomers of D-glucose. There are three classes of cyclodextrins-α,β, and γ-distinguished by the number of units of 1-4 linked glucoses. β-Cyclodextrin can function as a phase transfer agent in the nucleophilic displacement of 1-bromooctane by cyanide, iodide, and thiocyanate ion(13). It was found that β-cyclodextrin is also a useful phase transfer catalyst for the palladium chloride catalyzed oxidation of terminal and internal olefins(14). Of particular note is the oxidation of styrene to acetophenone in 80% yield using β-cyclodextrin/$PdCl_2$/$CuCl_2$/H_2O/O_2, a result which contrasts with carbon-carbon bond cleavage (i.e. benzaldehyde formation) when PEG-400 or a quaternary ammonium salt was used as the phase transfer catalyst.

Table II   β-Cyclodextrin and Palladium Chloride Catalyzed

Oxidation of Olefins

Substrate	Product	Yield (%)
1-butene	butanone	68
cis-2-butene	butanone	76
trans-2-butene	butanone	70
1-decene	2-decanone	61
	isomeric decenes	39
1,9-decadiene	2,9-decanedione	100
1,8-nonadiene	2,8-nonanedione	56
	1,8-nonanedione	7
allylbenzene	1-phenyl-2-propanone	17
	1-phenyl-1-propene	83
styrene	acetophenone	80
	benzaldehyde	10
	phenylacetaldehyde	10

Carbonylation

One of the most useful classes of metal and phase transfer catalyzed reactions are carbonylation reactions. Cobalt carbonyl is a valuable catalyst for such processes(15). When used in conjunction with methyl iodide, acetylcobalt carbonyl [$CH_3COCo(CO)_4$] is generated and can undergo addition to various unsaturated substrates including alkynes and Schiff bases. In addition, one can add this species to styrene oxides to give the enol

$$\text{Ar}\underset{O}{\overset{R}{\triangle}} + CO + CH_3I \xrightarrow[\text{NaOH, } C_6H_6]{Co_2(CO)_8, \text{CTAB}} \underset{HO}{\overset{Ar\underset{R}{}}{\underset{O}{\bigcirc}}}O \quad (5)$$

tautomer of furandiones. The driving force for this novel double carbonylation reaction is the apparent formation of an enol-cobalt complex(16).

Butenolides are formed in the alkyne-CH_3I-$Co_2(CO)_8$ phase transfer reaction. When the latter process is effected in the presence of ruthenium carbonyl, a second metal catalyst, γ-keto acids are isolated in good yields(17).

$$RC\!\!=\!\!CH + CH_3I + CO \xrightarrow[C_{12}H_{25}N(CH_3)_3{}^+Cl^-, C_6H_6, r.t.]{Co_2(CO)_8, Ru_3(CO)_{12}, NaOH} R\underset{COOH}{\overset{|}{C}}HCH_2COCH_3 \quad (6)$$

This is an authentic bimetallic phase transfer reaction in which the second metal species intercepts an organocobalt intermediate.

Complexes of other metals are also capable of catalyzing useful carbonylation reactions under phase transfer conditions. For example, certain palladium(o) catalysts, like $Co_2(CO)_8$, can catalyze the carbonylation of benzylic halides to carboxylic acids. When applied to vinylic dibromides, unsaturated diacids or diynes were obtained, using Pd(diphos)$_2$[diphos=1,2-bis(diphenylphosphino)ethane] as the metal catalyst, benzyltriethylammonium chloride as the phase transfer agent, and t-amyl alcohol or benzene as the organic phase(18). If, however, PEG-400 is employed as the solvent and phase transfer catalyst, under a nitrogen atmosphere, then the monoacid is obtained in good yield. Since vinylic dibromides are easily synthesized from carbonyl compounds, this constitutes a valuable method for oxidative homologation(19).

$$\underset{R'}{\overset{|}{R}C}\!\!=\!\!CBr_2 + CO \xrightarrow[PhCH_2N(C_2H_5)_3{}^+Cl^-, t\text{-AmOH}]{Pd(diphos)_2, 5N\ NaOH} \underset{R'}{\overset{|}{R}C}\!\!=\!\!C(COOH)_2 \quad (7)$$

$$RCOR' \xrightarrow[CBr_4]{Ph_3P} \underset{R'}{\overset{|}{R}C}\!\!=\!\!CBr_2 \xrightarrow[\substack{Pd(diphos)_2, N_2 \\ PEG\text{-}400 \\ 60\text{-}65°}]{NaOH\ or\ KOH} \underset{R'}{\overset{|}{R}C}HCOOH \quad (8)$$

Allyl chlorides and bromides are readily carbonylated to unsaturated acids using nickel cyanide and phase transfer catalysis conditions. Mechanistic studies revealed that the key catalytic species in this reaction is the cyanotricarbonylnickelate ion(20).

$$PhCH=CHCH_2Cl + CO \xrightarrow[\substack{(C_4H_9)_4N^+HSO_4^- \\ CH_3COCH_2CH(CH_3)_2}]{\substack{Ni(CN)_2 \\ 5M\,NaOH}} PhCH=CHCH_2COOH \quad 84\% \quad (9)$$

The latter may prove to be a valuable catalyst for a variety of phase transfer reactions.

Finally, manganese carbonyl complexes also show potential for effecting interesting phase transfer catalyzed carbonylation reactions. Alkynes react with carbon monoxide and methyl iodide in methylene chloride, using 5N NaOH as the aqueous phase, benzyltriethylammonium chloride as the phase transfer catalyst, and either bromopentacarbonylmanganese or dimanganese decacarbonyl to afford saturated lactones in good yields(21).

$$PhC\equiv CH + CH_3I + CO \xrightarrow[\substack{PhCH_2N(C_2H_5)_3^+Cl^- \\ 35°C,\ 1\ atm.}]{Mn(CO)_5Br,\ 5N\ NaOH} \quad \text{(lactone structure)} \quad (10)$$

47/31: trans/cis

In conclusion, phase transfer catalysis is a method of considerable potential for metal complex catalyzed reduction, oxidation and carbonylation reactions.

Acknowledgments

The above described results are due to the first-rate contributions of my co-workers who are cited in the references. I am indebted to the Natural Sciences and Engineering Research Council, and to British Petroleum, for support of this research.

References

1. Alper, H.; Amaratunga, S. Tetrahedron Lett. 1980, 2603.
2. Januszkiewicz, K.; Alper, H. J. Mol. Catal., 1983, 19, 139.
3. Hashem, K.E.; Petrignani, J.F.; Alper, h. J. Mol. Catal., 1984, 26, 285.
4. Joo, F.; Alper, H. Can. J. Chem. 1985, 63, 1157.
5. Alper, H.; Sibtain, F.; Heveling J. Tetrahedron Lett. 1983, 24, 5329.
6. Alper, H.; Heveling, J. J. Chem. Soc., Chem. Commun. 1983, 365.

7. Alper, H.; Sibtain, F. J. Organometal. Chem. 1985, 285, 225.
8. Januszkiewicz, K.; Alper, H. Organometallics, 1983, 2, 1055.
9. Januszkiewicz, K.; Alper, H. Tetrahedron Lett. 1983, 24, 5159.
10. Januszkiewicz, K.; Alper, H. Tetrahedron Lett. 1983, 24, 5163.
11. e.g., Gokel, G.W.; Goli, D.M.; Schultz, R.A. J. Org. Chem., 1983, 48, 2837; Neumann, R.; Sasson, Y. J. Org. Chem. 1984, 49, 3448.
12. Alper, H.; Januszkiewicz, K.; Smith, D.J.H. Tetrahedron Lett. 1985, 26, 2263.
13. Trifonov, A.Z.; Nikiforov, T.T. J. Mol. Catal. 1984, 24, 15.
14. Zahalka, H.A.; Januszkiewicz, K.; Alper, H. J. Mol. Catal. 1986, 35, 249.
15. Alper, H., Fund. Res. Homogen. Cat. 1984, 4, 79.
16. Alper, H.; Arzoumanian, H.;Petrignani, J.F.; Saldana-Maldonado, M. J. Chem. Soc., Chem. Commun. 1985, 340.
17. Alper, H.; Petrignani, J.F. J. Chem. Soc., Chem. Commun. 1983, 1154.
18. Galamb,V.; Gopal, M.; Alper, H. Organometallics, 1983, 2, 801.
19. Li, P.; Alper, H. unpublished results.
20. Joo, F.; Alper, H. Organometallics, 1985, 4, 1775.
21. Wang. J.X.; Alper, H., J. Org. Chem. 1986, 51, 273.

RECEIVED August 20, 1986

Chapter 3

Mechanism of Phase-Transfer Catalysis: The Omega Phase

Charles L. Liotta, Edward M. Burgess, Charles C. Ray, Elzie D. Black, and Barbara E. Fair

School of Chemistry, Georgia Institute of Technology, Atlanta, GA 30332

The use of polyethers and quaternary salts as liquid-liquid and solid-liquid phase transfer catalysts has been well-documented in the literature.[1,2,3] It has been shown that (1) the catalyst functions as a vehicle for transferring the anion of a metal salt from the aqueous or solid phase into the organic phase where reaction with an organic substrate ensues, (2) the rate of reaction is proportional to the concentration of the catalyst in the organic phase,[4] and (3) small quantities of water have a significant effect on the catalytic process.[5,6,7] This Communication specifically addresses the role of cyclic polyethers as phase transfer catalysts and the influence of water with regard to the location of the catalyst.

The rates of reaction of benzyl bromide and benzyl chloride with potassium cyanide were studied as a function of added water, in the presence and absence of 18-crown-6 (Equation 1). These heterogeneous reactions were carried out in toluene (50 mL) at 85°C and 25°C.[8]

$$\text{C}_6\text{H}_5\text{CH}_2\text{X} + \text{K}^+\text{CN}^- \xrightarrow[\text{Toluene}]{\text{18-Crown-6}} \text{C}_6\text{H}_5\text{CH}_2\text{CN} + \text{K}^+\text{X}^- \quad (1)$$

The data for the reactions of potassium cyanide with benzyl halides at 85°C and 25°C are summarized in Tables I-III and graphical representations of these data are shown in Figures 1-3. The reactions carried out at 85°C were followed to 70% completion, while those at 25°C were followed to 50% completion. In general, excellent first-order kinetic plots were obtained. Each point on the graphs represents an average of at least three kinetic determinations. It is interesting to note that in the absence of added water (solid-liquid phase transfer catalysis), the rates of benzyl halide disappearance were more accurately described by zero-order kinetics.

The data in Tables I-III clearly indicate that the amount of added water had a marked effect on the rate of reaction. In the presence of 18-crown-6, addition of minute quantities of water caused a dramatic increase in rate. Beyond this maximum, as the quantity of

Table I. The Rate of Reaction of Benzyl Bromide with Potassium Cyanide at 85°C in the Presence and Absence of 18-Crown-6 as a Function of Added Water

Volume of Water (mL)	$k \times 10^4 \text{ sec}^{-1}$ (a)	$k \times 10^4 \text{ sec}^{-1}$ (b)	$k_{cat} \times 10^4 \text{ sec}^{-1}$
0	0.02^c	0.0	0.02
1	13.9	0.0	13.9
10	11.9	0.0	11.9
20	6.8	0.6	6.2
30	4.4	1.0	3.4
40	3.2	1.3	1.9
50	2.8	2.1	0.7

a. Conditions: 0.05 mol benzyl bromide, 0.0025 mol 18-crown-6, 0.15 mol KBr, 0.15 mol KCN.

b. 18-Crown-6 omitted.

c. Zero order (M sec^{-1})

Table II. The Rate of Reaction of Benzyl Chloride with Potassium Cyanide at 85°C in the Presence and Absence of 18-Crown-6 as a Function of Added Water

Volume of Water (mL)	$k \times 10^5$ sec^{-1} (a)	$k \times 19^5$ sec^{-1} (b)	$k_{cat} \times 10^5$ sec^{-1}
0.0	3.2^c	0.0	3.2
0.36	9.2	0.0	9.2
0.50	9.4	0.0	9.4
1.0	11.6	0.0	11.6
2.0	14.7	0.0	14.7
10.0	10.2	0.0	10.2
15.0	6.9	0.7	6.2
20.0	5.8	1.3	4.5
30.0	5.0	1.7	3.3
40.0	3.9	1.9	2.0
50.0	4.2	2.5	1.7
75.0	4.8	3.2	1.6

a. Conditions: 0.05 mol benzyl chloride, 0.010 mol 18-crown-6, 0.15 mol KCl, 0.15 mol KCN.

b) 18-Crown-6 omitted.

c) Zero order (M sec^{-1})

Table III. The Rate of Reaction of Benzyl Bromide with Potassium Cyanide at 25°C in the Presence of 18-Crown-6 as a Function of Added Water

Volume of Water (mL)	$k \times 10^5 \text{ sec}^{-1}$ [a]
0.0	2.06[b]
0.5	1.71
1.2	2.54
1.4	2.84
1.5	9.40
1.6	5.10
1.75	5.34
2.0	5.12
2.5	3.36
3.0	3.50
3.5	3.26
4.0	2.20
6.0	1.49

a. Conditions: 0.05 mol benzyl bromide, 0.010 mol 18-crown 6, 0.15 mol KBr, 0.15 mol KCN.

b. Zero order ($M \text{ sec}^{-1}$)

water increased, the first order rate constant decreased.[9] It is important to note that in the absence of crown, no reaction took place during the same time period in the presence of less than ca. 10 mL of water. With the addition of larger quantities of water, hydrolysis of the benzyl halides becomes a measurable process in both the presence and absence of crown. In order to eliminate this uninteresting pathway from the analysis k_{cat} was defined as the difference in the rate constants between the crown-catalyzed reactions and those reactions with no crown present. The values for k_{cat} are also tabulated in Tables I and II. The studies conducted at 25°C (Table III) employed only small quantities of water where the uncatalyzed (absence of 18-crown-6) reaction rates were negligibly slow.

In order to address the relationship between the rate profiles (Figures 1, 2 and 3) and the location of the 18-crown-6, the following series of experiments were conducted. 18-Crown-6 (0.004 mol) was dissolved in 10 mL of toluene containing a mixture of potassium chloride and potassium cyanide (0.027 mol each). Varying quantities of water were added. At each addition, the system was allowed to equilibrate at room temperature and the quantity of 18-crown-6 in the organic phase was measured by capillary gas chromatography. The results are summarized in Table IV and a graphical representation is shown in Figure 4. Initially, all the crown was present in the toluene. Addition of the salt mixture produced a slight decrease in the amount of crown in the organic phase. Quite surprisingly, addition of minute quantities of water (totalling 0.08 mL) resulted in a 97% depletion of the catalyst from the organic phase. That the 18-crown-6 was translocated onto the surface of the salt was proven by filtering the solid salt, drying it, and extracting it with methylene chloride. All of the 18-crown-6 was recovered.[10]

It is evident from Figures 1-3 that the greatest catalytic activity observed in the reaction of the benzyl halides with cyanide ion takes place at very low concentrations of water. In conjunction with this, Figure 4 indicates that only minute quantities of the catalyst is present in the organic phase (<3%); the catalyst is primarily located on the surface of the salt. It is conjectured that the initial water added to the system coats the surface of the salt and it is this aqueous salt coating which extracts the crown from the organic phase. This new region of the reaction system, the omega phase, appears to be intimately in the catalytic reaction process.

It has already been mentioned that in the absence of added water the reaction kinetics follows a pseudo zero-order rate profile; the rate-controlling step under these conditions appears to involve the complexation of the crown in the organic phase with the salt in the solid phase. In contrast to this, in the presence of small quantities of water the reaction kinetics follows a pseudo first-order rate profile. Thus it appears that the water facilitates the interaction between the crown and the salt by forming an omega phase since the displacement process now becomes the rate-controlling step. The phase region where the displacement process actually takes place is not certain at this juncture.

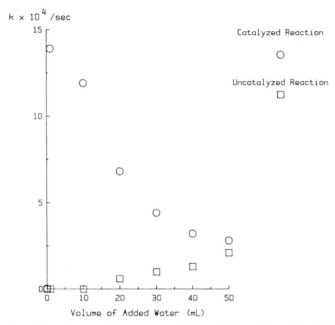

Figure 1. The Rate of Reaction of Benzyl Bromide with KCN at 85°C in the Presence and Absence of 18-Crown-6 as a Function of Water.

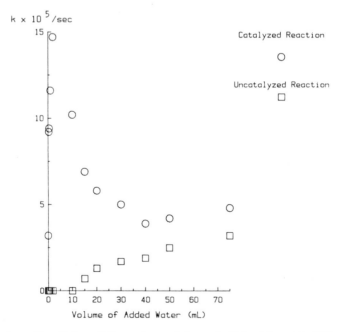

Figure 2. The Rate of Reaction of Benzyl Chloride with KCN at 85°C in the Presence and Absence of 18-Crown-6 as a Function of Water.

Table IV. The Effect of Added Water on the Concentration of 18-Crown-6 in Toluene at Ambient Temperature

Volume of Water (L)	Equivalents[b] of Water	% Crown[a] in Toluene
0	0.00	91.5
10	0.14	81.4
15	0.21	77.3
21	0.29	72.7
22	0.31	50.0
23	0.32	40.0
25	0.35	34.6
30	0.42	17.7
36	0.50	5.8
45	1.25	2.5
50	1.39	2.0
80	2.22	0.0

a. Conditions: 0.0040 mol 18-crown-6, 0.027 mol KCl, 0.027 mol KCN, 10 mL toluene.

b. Moles of H_2O/moles of crown

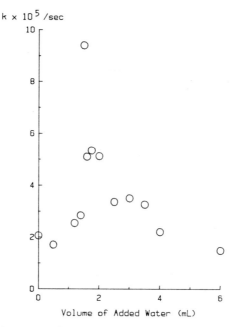

Figure 3. The Rate of Reaction of Benzyl Bromide with KCN at 25°C in the Presence of 18-Crown-6 as a Function of Water.

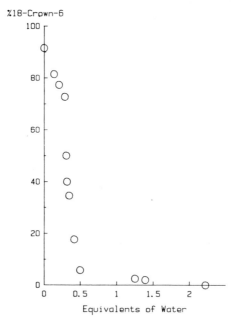

Figure 4. The Effect of Added Water on the Concentration of 18-Crown-6 in Toluene at Ambient Temperature.

Omega-phase formation has been found to occur with a variety of catalysts and salts. Indeed, initial experiments have demonstrated that triglyme and tetraglyme are extracted from the organic (toluene or benzene) phase onto the surface of inorganic salts upon small quantities of water. Similarly, a range of alkali and alkaline earth metal salts including KSCN, KF, CH_3COOK, LiI, NaBr, $MgCl_2$, and $CaCl_2$ have been shown to participate in omega-phase formation. The nature and catalytic activity of these systems is presently under investigation.

References

1. Starks, C.M.; Liotta, C.L. "Phase Transfer Catalysis: Principles and Practice"; Academic Press: New York, 1978.
2. Dehmlow, E.V.; Dehmlow, S.S. "Phase Transfer Catalysis"; 2nd Edn. Verlag Chemie: Weinheim, 1983.
3. Weber, W.P.; Gokel, G.W. "Phase Transfer Catalysis in Organic Synthesis"; Springer-Verlag: Berlin, Heidelberg, New York, 1977.
4. Starks, C.M.; Owens, R.M. J. Am. Chem. Soc. **1973**, 95, 3613.
5. Zahalka, H.A.; Sasson, Y. J. Chem. Soc. Chem. Commun. **1984**, 1652; and references therein.
6. Gerbi, D.J.; Dimotsis, G.; Morgan, J.L; Williams, R.F., J. of Polymer Sci., **1985**, 23, 551.
7. Delmas, M.; Le Bigot, Y.; Gaset, A.; Tetrahedron Letters, **1980**, Vol. 21, 4831.
8. Rigorously anhydrous materials were employed to ensure control of the quantity of water present in each kinetic experiment. Reactions at the elevated temperature were performed in standard three-necked flasks equipped with a reflux condenser and nitrogen adapter. These reaction mixtures were mechanically stirred (1550 rpm) and heated in an oil bath thermostated at 85.0±.5°C with an I^2R Therm-o-watch. The room temperature studies were performed in one-neck flasks employing an excursion mixer (230 excursions/min, 10 cm strokes) at room temperature (25±1°C). In both studies, the disappearance of benzyl halide was monitored by capillary gas chromatographic analysis of samples which were removed from the organic phase. The benzyl halide concentration was determined by comparison with an internal standard, utilizing the internal standard method of data analysis of the Hewlett-Packard 3390A integrator.
9. A similar dependence of the first-order rate constants with respect to the quantity of added water has been reported for the reaction of sodium formate with 1,4-dichlorobutane and related displacement reactions.[5] In these studies tetra-n-butylammonium hydrogen sulphate and tetra-n-butylammonium bromide were used as catalysts and chlorobenzene as the solvent.
10. The use of dicyclohexyl 18-crown-6 (in place of 18-crown-6) produced the same rate profile with respect to the quantity of added water. It was found that 30% of the decyclohexyl 18-crown-6 remained in the organic phase upon addition of 0.25 ml of water; 70% of the crown was associated with the water-salt phase.

RECEIVED July 26, 1986

Chapter 4

Cation-Binding Properties of Crown Ethers, Lariat Ethers, Bibracchial Lariat Ethers, and Poly(ethylene glycols) as Potential Phase-Transfer Catalysts

George W. Gokel, K. Arnold, T. Cleary, R. Friese, V. Gatto, D. Goli, C. Hanlon, M. Kim, S. Miller, M. Ouchi, I. Posey, A. Sandler, A. Viscariello, B. White, J. Wolfe, and H. Yoo

Department of Chemistry, University of Miami, Coral Gables, FL 33124

> During recent years, there has been considerable interest in both linear open-chained and cyclic polyether compounds as phase transfer catalysts. According to the much discussed "hole size relationship," small crowns favor binding to small cations and larger, often more readily available crowns are thought to be ineffective. Non-cyclic polyethers cannot exhibit hole-size selectivity and are thus thought to be poor phase transfer catalysts. The binding of various polyether species, including polyethylene glycols, crown ethers, lariat ethers, bibracchial lariat ethers, is compared and evidence is presented to help understand some aspects of binding strengths, selectivities, and the mechanism of catalysis in certain cases.

Phase transfer catalytic processes (1-3) have been the subject of intensive study in many laboratories throughout the world since its potential was recognized almost simultaneously and independently by Starks (4) and Makosza (5). The principles outlined by Starks in 1971 (6) have generally stood the test of time even though many compounds besides quaternary 'onium salts have been utilized as phase transfer catalysts (1-3).

'Onium Salt Catalysts

The first catalysts utilized in phase transfer processes were quaternary 'onium salts. In particular, benzyltriethylammonium chloride was favored by Makosza (7) whereas Starks utilized the more thermally stable phosphonium salts (6,8). In either case, the catalytic process worked in the same way: the ammonium or phosphonium cation exchanged for the cation associated with the nucleophilic reagent salt. The new reagent, Q^+Nu^-, dissolved in the organic phase and effected substitution.

$$Q^+Nu^- + R-X \longrightarrow R-Nu + Q^+X^-$$

The advantages of the quaternary 'onium ion pair was twofold.

0097-6156/87/0326-0024$06.00/0
© 1987 American Chemical Society

First, it permitted the anion Nu^- to dissolve in an organic phase when the inorganic salt, M^+Nu^-, proved insoluble. Second, since the principal mode of solvation in the organic phase is lipophilic-lipophilic interactions, the cation (Q^+) is well solvated but the counteranion (Nu^-) is not. The reactivity of Q^+Nu^- is therefore greater than the reactivity of M^+Nu^- although completely valid comparisons are limited by solubility.

Crown Ethers as Phase Transfer Catalysts

Almost from the first disclosures of phase transfer catalytic processes, alternate catalysts have been sought which it was hoped would be more selective, less expensive, more stable, etc. Notable among these alternative catalysts were the macrocyclic "crown" polyethers. More thermally stable and less prone to undergo degradative elimination than ammonium salts, they are also more expensive than many catalysts which are their equal in efficacy at lower temperatures.

One supposed disadvantage of crowns as phase transfer catalysts was the so-called "hole size relationship." It has generally been thought that 15-crown-5, whose hole size is similar to sodium's cation diameter, was selective for Na^+ over other cations. Likewise, the selectivity of 18-crown-6 for K^+ made it the better choice for reactions involving that cation. Since sodium salts are generally cheaper and always of lower molecular weight than the analogous potassium salts, they are usually favored for large-scale processes. Of the simple crowns, however, the cheapest ones are all 18-membered rings. Aldrich Chemical Company (9), for example, sells 25 g lots of 12-crown-4, 15-crown-5, and 18-crown-6 for $44.10, $40.00, and $33.50 respectively. 18-Crown-6 has historically been the most accessible crown ether (10). Because it was thought that 18-crown-6 would bind K^+ and not Na^+, much work involving crowns as catalysts for industrial processes was de-emphasized. We note parenthetically that this lack of enthusiasm was also based on the widespread belief that crowns are very toxic. The available evidence certainly does not support this presumed toxicity (11-21).

Evidence Concerning the "Hole-Size Relationship"

The binding process is, to coin a phrase, complex. In homogeneous solution, it can be expressed by the equation

$$M^+ + L \xrightleftharpoons{K_S} [M \cdot L]^+$$

in which M^+ is a cation and L is a ligand. The equilibrium constant for this process, K_S, is called the stability or binding constant and is equal to $k_{complex}/k_{decomplex}$ or simply $K_S = k_1/k_{-1}$. When the aqueous/organic phase equilibria are considered, the process becomes even more complicated. In an effort to understand the binding process, we measured the Na^+, K^+, Ca^{2+}, and NH_4^+ stability constants in anhydrous methanol solution with 12-crown-4, 15-crown-5, 18-crown-6, 21-crown-7, and 24-crown-8 (22). The results are shown in Table I.

Table I. Stability Constants in Methanol at 25 °C

Crown	Na$^+$	Log K_S in MeOH K$^+$	NH$_4^+$	Ca^{2+}
12-crown-4	1.7	1.74	1.3	----
15-crown-5	3.24	3.43	3.03	2.36
18-crown-6	4.35	6.08	4.14	3.90
21-crown-7	2.54	4.35	3.27	2.80
24-crown-8	2.35	3.53	2.63	2.66

If the notion that 15-crown-5 is selective for Na$^+$ is correct, then K_S for the interaction 15-crown-5·Na$^+$ should be greater than the interaction 15-crown-5·K$^+$. We see from Table I that log K_S for the sodium interaction is 3.24 (1,740) compared to 3.43 (2,690) for potassium. An alternate definition of Na$^+$-selectivity might be that 15-crown-5 binds Na$^+$ more strongly than 18-crown-6 binds Na$^+$. The log K_S values for this pair of interactions is, respectively, 3.24 (1,740) vs. 4.35 (22,400). Clearly, 15-crown-5 is not selective for Na$^+$ and the simple "hole-size rule" does not correctly account for the results.

A perusal of the data in Table I reveals that for the series of cations surveyed, two rules can be formulated. (i) All of the crowns in this series bind K$^+$ more strongly than any of the three other cations. (ii) 18-Crown-6 is the best cation binder in this group of five ligands regardless of which cation is considered. From these data, it is clear that the less expensive 18-crown-6 would not only be acceptable for phase transfer reactions involving Na$^+$ salts, it would be preferable.

Many factors affect cation binding selectivities, not all of which are well understood. We offer two suggestions which help rationalize these observations. The special complexing strength of 18-crown-6 can be understood as follows. First, six oxygen donor groups are available. This is important to provide a minimum level of cation solvation. Second, there is neither front nor back strain in the macroring since it has an even number of -CH$_2$-CH$_2$-O- groups and it is optimally sized. Third, when complexed, the macroring often exhibits the favorable D$_{3d}$ symmetry. The selectivity for K$^+$ over the other cations can be understood in terms of solvation enthalpy. Of the cations studied, K$^+$ has the lowest solvation enthalpy. The crown can compete more effectively with solvent when the solvation forces are weakest. We offer these observations to assist the reader in considering the problem of selectivity. Others (23-25) have studied the cation complexation problem in some detail.

The lesson from these studies is that selectivity is not necessarily controlled by a similarity in size of cation and macrocycle hole. In addition, crowns of all sizes show at least some affinity for cations of different sizes (26). One can therefore consider using a convenient crown as catalyst rather than seeking a specific one which might be only marginally better or, indeed, might be inferior.

Oligoethylene Glycols as Phase Transfer Catalysts

Oligoethylene glycols are the open-chained analogs of crown ethers. They contain the same essential features as crowns, except that they are non-cyclic. In principle, such compounds could catalyze phase transfer reactions although they might be less effective. The reduced efficacy might be predicted based on an entropy effect, i.e. that the energy cost of wrapping about a cation overwhelms the payback in cation stabilization. In addition, simple oligoethylene glycols contain terminal $-CH_2-CH_2-OH$ units and might undergo elimination more readily than the crowns. On the other hand, should the oligoethylene glycols catalyze phase transfer reactions, these difficulties would be more than compensated by the dramatically lower cost. Compared to the 18-crown-6 (MW 264) per gram cost of \$1.34, poly(ethylene glycol) of average MW 300, costs \$0.018 per gram (9). The nearly 100-fold price differential is a considerable incentive for investigating the properties of these "crown-like" compounds. In fact, these inexpensive polymers catalyze phase transfer reactions quite effectively (27).

Two interesting questions arise concerning poly(ethylene glycols) and crowns. First, what are the binding strengths of various PEGs with respect to Na^+ or other cations? Second, are higher molecular weight PEGs (PEG-10000 for example) more or less effective phase transfer catalysts than lower molecular weight compounds (eg. PEG-300)? A third, and obviously intriguing, question is whether catalysis is efficient and if so, does it occur by the same mechanism as crown catalysis?

The first question was answered by determining log K_S values for Na^+ in anhydrous methanol solution with poly(ethylene glycol)s in the molecular weight range 200 to 14,000. The stability constants (log K_S) ranged from 1.64 (44) to 4.08 (12,000). A plot of log K_S vs. log molecular weight is a straight line with a slope of about 1.4 (27). The stability constants for the corresponding poly(ethylene glycol) monomethyl ethers are slightly lower, but the lines are nearly parallel. In either case, the lines are straight. Since there is no peak nor any ultimate decrease in binding strength, the notion of "size-selective wrapping" cannot be true (28).

Each poly(ethylene glycol) has many oxygen donor groups. The more Lewis basic donors which are present in each molecule, the stronger the cation binding. Experimentally, binding is measured by the conductivity of cation-containing solutions in the presence and absence of ligand. When the potentials of equal weights (1 g each) of PEG-500 and PEG-14000 were measured, the voltage readings were the same. In other words, the binding is a function of the total number of binding sites present and not the number of polymer chains. This suggests that a long chain may be involved in binding more than one cation.

Is the phase transfer catalytic ability of these materials independent of molecular weight as the cation binding is? The answer to this question was determined kinetically (27). The rate of the simple, S_N2 reaction shown below was monitored in the presence of different catalysts and different amounts of oligoethylene glycols.

$$C_8H_{17}-Cl + NaCN \xrightarrow[\text{water and decane}]{\text{catalyst}} C_8H_{17}-CN + NaCl$$

The reaction was originally used by Starks (6) in his early studies of ptc mechanism. The reaction was monitored by gas-liquid chromatography and decane served both as solvent and internal standard. The reaction was run using a quaternary 'onium salt, Aliquat 336, and 18-crown-6 for calibration. The rates are shown in Table II. Please note that the relative rate of 1.00 corresponds to an absolute rate at 105 °C of 1.34×10^{-6} $M^{-1}sec^{-1}$.

Table II. Catalyst Dependence of Rate in PTC

Catalyst	Amt, g (mol-%)	Rel. Rate
None	0 (0)	0
Aliquat 336	0.404 (1.5)	269
18-Crown-6	0.264 (1.5)	0.8
PEG-400	0.400 (1.5)	1.00
PEG-3400	3.400 (1.5)	1.5
PEG-MME-750	0.750 (1.5)	0.3
18-Crown-6	3.400 (19.3)	27
PEG-400	3.400 (12.75)	18.7
PEG-3400	3.400 (1.5)	1.5
PEG-MME-750	3.400 (6.8)	2.5

Several conclusions can be reached from these data. The first is that on a per mole basis, the quaternary ammonium salt is the most favorable catalyst for this reaction. Among the other compounds, the catalytic activity of 1.5 mole-% of crown, PEG, or PEG-MME are similar. If equal weights of PEG-400 and PEG-3400 are used, quite different reaction rates are observed. This is because each polymer chain is capable of transporting one cation across the phase boundary at a time. The ratio of molecular weights is 8.5, so there are 8.5 more catalysts available in the PEG-400 catalyzed reaction than in the one involving the higher molecular weight compound. The actual ratio of rates for these two processes is 12.5, or nearly the expected value.

A one ligand-one cation mechanism seems entirely reasonable. Transport of any cation into a lipophilic medium requires lipophilic solvation. The PEGs are compounds of intermediate polarity. They contain many $-CH_2-CH_2-$ groups, but they also contain an equal number of polar oxygen atoms. A solvent like decane can apparently solvate a cation-complexed PEG when only one charge is present, but not when there are two or more charges associated with the chain. The mechanism of PEG catalysis is therefore similar to the mechanism of crown catalysis, but there is more "wasted" catalyst in longer-chain PEGs.

The Lariat Ethers

Although the lariat ethers (29-31) were conceived on principles related to biological activity, they are interesting candidates for study as either free phase transfer catalysts, or as polymer-bound catalysts. In the latter case, the sidearm could serve both a complexing function and as a mechanical link between macroring and polymer. Polymeric phase transfer catalyst systems have been prepared previously and fall into five general groups. These are: polysulfoxides (32-34), polymer-supported or based quaternary 'onium salts (35-41), polymer-supported PEGs (42-45); polymer-supported crowns and polycrowns (46-50), and miscellaneous supported reagents including those on clay (51), alumina (52), and metal oxide (53) supports. The basic structural type envisioned for the lariat ethers is shown below.

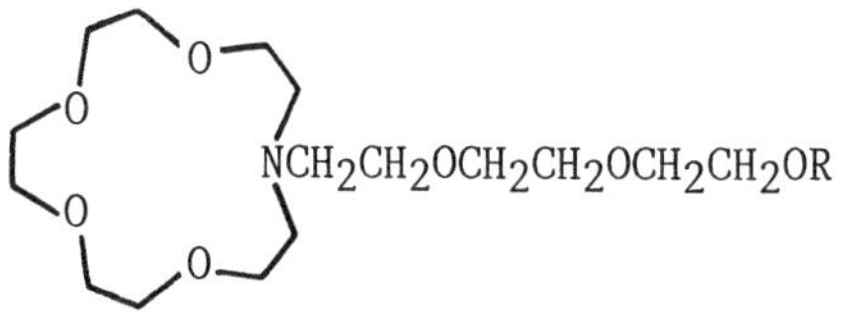

NCH$_2$CH$_2$OCH$_2$CH$_2$OCH$_2$CH$_2$OR

R = CH$_3$ (free ligand)

R = H (possible polymer link)

The structure shown above is a representative of the class of lariat ethers we have called "nitrogen-pivot" compounds (31). The presence of invertable nitrogen as the point of sidearm attachment makes these compounds more flexible than the corresponding "carbon-pivot" lariat ethers (29,30). The carbon- and nitrogen-pivot lariat ethers have been prepared with a variety of sidechains attached. Examples of compounds which could be linked to existing polymers such as chloromethylated polystyrene or copolymerized with other monomers are shown below.

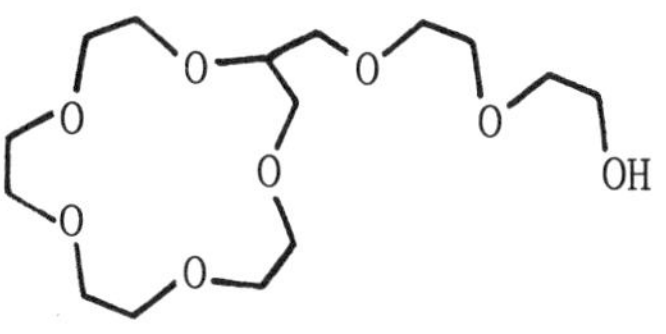

N-CH$_2$-CH=CH$_2$

Carbon-pivot 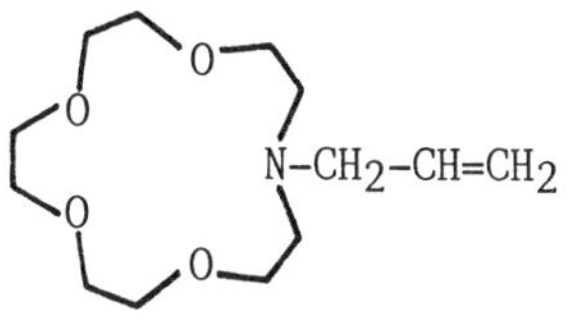   Nitrogen-pivot

We believe that the nitrogen-pivot compounds are more flexible than the carbon-pivot structures. This conclusion is based on cation binding data and on solution studies involving ^{13}C-NMR relaxation time and other NMR techniques (54,55). In part as a result of this, the cation binding affinities for the nitrogen-pivot lariats (31) generally exceed those of the carbon-pivot (29,30) compounds.

Sodium Cation Binding by Nitrogen-pivot Lariat Ethers

Once the broad range of closely related nitrogen-pivot macrocycles was in hand, we wished to assess the sodium cation binding affinities of these species. Such studies should give insight into, among others, such questions as (i) ring-size effects; (ii) macroring-sidearm cooperativity, and (iii) donor group effects.

The homogeneous cation binding constants, i.e., the equilibrium constant (K_S) for the equation

$$Na^+ + lariat\ ether = complex^+$$

were measured using ion selective electrode methods (31). The results were plotted as log K_S for Na^+ in anhydrous methanol solution vs. the total number of oxygen donor groups in the lariat (see Figure 1). The striking feature of this graph is that when the number of oxygen donor groups is identical, the binding is identical irrespective of the donor group distribution. This suggests that in a flexible system, the precise arrangement of donor groups is far less important than the presence of a certain number of them. The two six-oxygen systems, N-[(2-methoxyethoxy)ethyl]monoaza-15-crown-5 and N-(2-methoxyethyl)monoaza-18-crown-6, both bind Na^+ more strongly than does 18-crown-6. This is an encouraging observation since 18-crown-6 has found frequent application in preparative ptc.

Methods are now available for the synthesis of N-substituted monoaza-crowns with $>N(CH_2CH_2O)_n-H$ sidearms (31). Such molecules could be attached to chloromethylated polystyrene as indicated in references 35-50.

Intramolecularity in Lariat Ether Complexes

For lariat ethers to be effective as polymer-bound phase transfer catalysts, sidearm and macroring cooperation must be intramolecular. It is unlikely that two lariat ethers will be close enough on a polymer backbone or other support for the ring of one compound to interact with the sidearm donors of another. The mechanical attributes of lariat ethers will be independent of spacing but for any advantage in cation binding and anion activation to be realized, the macroring and its attached sidearm must cooperate to envelop the cation, solvate it, and shield it from the counteranion.

Intramolecularity in lariat ether complexation was demonstrated in three ways. First, when solution cation binding constants were determined, K_S was shown to be independent of cation and macrocycle concentrations (within certain limits). The efficacy of complexation of a single cation by two macrorings [$(ML_2)^+$ complex formation] should be concentration dependent. Second, ammonium cation binding constants were determined for the series (see above) of monoaza-15-crown-5 and -18-crown-6 compounds having $(CH_2CH_2O)_nCH_3$ (n = 0 to 8) sidearms (31).

Ammonium cation binding differs from that of alkali metals because the former is a tetrahedral cation and the latter are

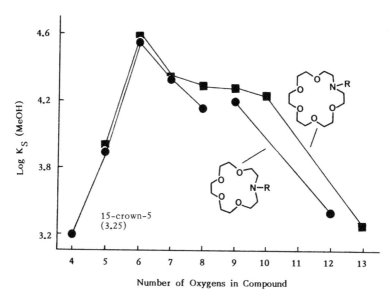

Figure 1. Plot of sodium cation binding in anhydrous methanol (at 25±0.1 °C) vs. the total number of oxygen atoms present in each ligand.

spherical. Ammonium ion forms three, well-space hydrogen bonds to an 18-membered crown ring and a single >N-H is perpendicular to the crown plane. A study of CPK molecular models led to the prediction that the second oxygen in a $CH_2CH_2OCH_2CH_2OCH_3$ sidearm would reach the perpendicular >NH bond. Since five oxygen atoms are present in a monoaza-18-crown-6 compound, a binding peak was predicted for NH_4^+ binding to occur when seven oxygen atoms are present. Due to the specificity of tetrahedral ammonium ion's H-bonds, a prediction of poor binding by all 15-membered rings could be made. Indeed, exactly these results were obtained (31,56) confirming solution phase intramolecularity.

Solid State Complex Structures

The third confirmation of intramolecularity came from X-ray crystal structure analysis of several lariat ethers (57-59). The crystal structures of K^+ complexes of N-(2-methoxyethyl)monoaza-15-crown-5 and -18-crown-6 (57) are shown in Figure 2. These structures are drawn by the computer to show only the donor groups. Each solid line which appears to be an O-O or N-O bond is actually a $-CH_2CH_2-$ link. Such drawings were first made at the suggestion of Prof. Richard Gandour to emphasize the overall donor group geometry of each complex. An artifact of such drawings is that -N< appears to be tetrahedral and to have its electron pair turned away from the cation. This is not the case in any of the structures determined to date.

Note that K^+ is slightly above the 18-membered ring's donor group plane but it is well outside the smaller macroring and fully enveloped by it. Obviously, a longer chain than four atoms would be required if the sidearm must secure the macroring to a polymer as well as help complex the cation. The only structure thus far in hand which bears on this question is the solid state structure of the monoaza-12-crown-4 derivative having a three-ethyleneoxy sidechain (60). Because the macroring is so small, all six oxygens are involved in cation binding and none of the chain is left for mechanical purposes. Whether this observation is extensible to systems having longer chains is currently unknown because complexes of lariat ethers having long sidechains have thus far proved impossible for us to crystallize.

Bibracchial Lariat Ethers: BiBLEs

One way to ensure that both a mechanical link and donor groups are available for cation binding is to utilize two arms. We use the Latin word bracchium (arm) in naming these structures and call them bibracchial lariat ethers. We abbreviate using the acronym "BiBLE" (60). Three simple BiBLEs are illustrated below. In each case, the synthesis is a simple, single-step reaction of a primary amine with triethyleneglycol diiodide and sodium carbonate in acetonitrile solution (60). The compounds having the first and second sidearms illustrated below were prepared from ethanolamine and methoxyethylamine respectively. The third compound was made by alkylation of diaza-18-crown-6 with ethyl bromoacetate.

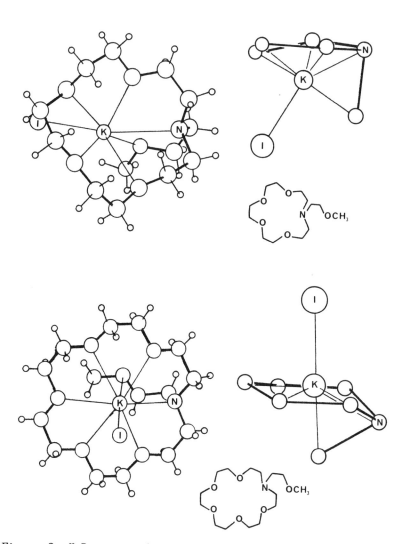

Figure 2. X-Ray crystal structures of KI complexing with N-(2-methoxyethyl)-monoaza-15-crown-5 and N-(2-methoxyethyl)-monoaza-18-crown-6. Complete structures are illustrated at the left and "framework" structures showing only KI and the ligand's donor groups are illustrated at the right. Structures were determined in collaboration with F. Fronczek and R. Gandour.

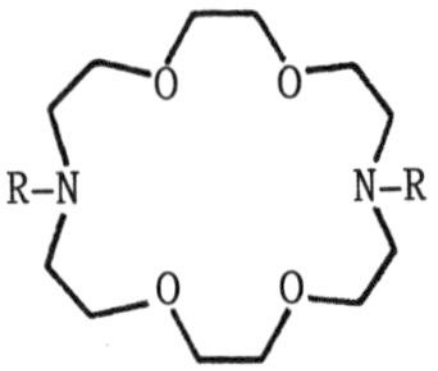

R = CH$_2$CH$_2$OH

R = CH$_2$CH$_2$OCH$_3$

R = CH$_2$-COO-C$_2$H$_5$

The cation binding strengths of these compounds proved substantial and a change in relative selectivity emerged for the more charge-dense cations Na$^+$ and Ca^{2+} over K$^+$ (<u>60</u>). For example, the three compounds shown above have alcohol, ether, and ester sidearms. The alcohol and ester sidearms show Na$^+$ <u>vs.</u> K$^+$ binding (log K$_S$) of 4.87/5.08 and 5.51/5.78, respectively. In contrast, when the sidearm is <u>N</u>-(2-methoxybenzyl), the Na$^+$/K$^+$ ratio is 3.65/4.94. For the alcohol and ester sidearms shown above, the Ca^{2+} binding strengths are 6.02 and 6.78, respectively. Such a high calcium-cation selectivity may be of little practical value in the conduct of phase transfer reactions, but it is nevertheless both interesting and striking.

In order to gain the maximum advantage in the application of these compounds, we sought a synthesis which was both efficient and flexible enough to permit preparation of 15-membered ring BiBLEs. Naturally, such a synthesis would also be of use in the synthesis of symmetrical, 18-membered ring BiBLEs.

Such an approach is shown below (<u>61</u>). A primary amine was treated with diglycoyl chloride to form the <u>bis</u>(amide). Reduction using LiAlH$_4$ or BH$_3$ afforded the bridged diamine in good yield. The diamine was then allowed to react with 1,2-<u>bis</u>(iodoethoxy)ethane (NaI, CH$_3$CN) to form the N,N'-disubstituted-4,10-diaza-15-crown-5 derivative. In this case, the crown-diether (>NCH$_2$CH$_2$OCH$_3$) exhibited a Na$^+$/K$^+$ ratio of 5.09/4.86. This compares with a ratio of 4.75/5.46 for the corresponding 18-membered ring diether (<u>60</u>).

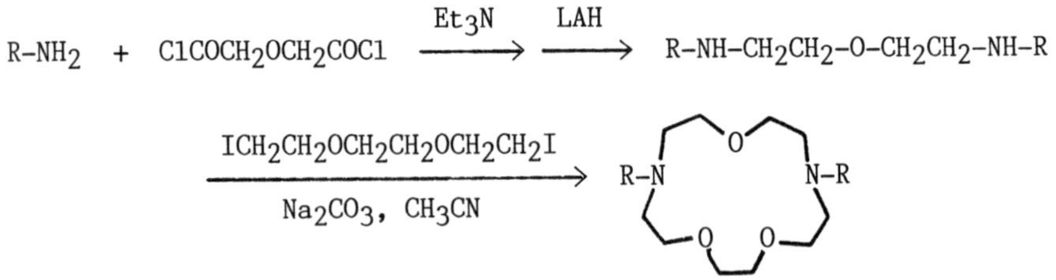

R-NH$_2$ + ClCOCH$_2$OCH$_2$COCl $\xrightarrow{\text{Et}_3\text{N}}$ $\xrightarrow{\text{LAH}}$ R-NH-CH$_2$CH$_2$-O-CH$_2$CH$_2$-NH-R

$\xrightarrow[\text{Na}_2\text{CO}_3, \text{CH}_3\text{CN}]{\text{ICH}_2\text{CH}_2\text{OCH}_2\text{CH}_2\text{OCH}_2\text{CH}_2\text{I}}$

Of course, phase transfer reactions are generally run using a single metal cation in the source phase. Achieving selectivity of Na$^+$ over K$^+$ is of less practical significance than enhancing the Na$^+$-binding strength. Phase transfer catalysts which are strong Na$^+$ complexers and have the advantage of being polymer-bound and readily recoverable, will surely be of interest.

Summary

Oligoethylene glycols are relatively weak cation binders but they are effective phase transfer catalysts. The combination of an oligoethylene glycol-like sidearm and a macroring leads to a class of compounds we call lariat ethers. We have now demonstrated that lariat ethers having one or two arms can be readily synthesized. These compounds not only bind cations, but they are often stronger cation complexers than simple crowns like 18-crown-6. The substances complex cations using both the macroring and sidearm of the same molecule and the sidearms can be used to mechanically link the crown to a polymer support. The lariat ethers can be tailored to achieve cation selectivity and binding strength for selected cations. Sodium cation-selectivity is especially attractive since Na^+ is a favored cation for salts used in phase transfer processes. In future work, we expect to demonstrate the practice of these studies by preparing novel polymer-bound lariat ethers and applying them to phase transfer processes.

Acknowledgments

We thank the National Institutes of Health and W. R. Grace & Co. for grants which supported portions of this work.

Literature Cited

*Address correspondence to this author at the University of Miami. Portions of this work were done at the University of Maryland.

1. Weber, W.P.; Gokel, G.W.; Phase Transfer Catalysis in Organic Synthesis, Springer Verlag, Berlin, 1977.
2. Starks, C.M.; Liotta, C.L.; Phase Transfer Catalysis Academic Press, New York, 1978.
3. Dehmlow, E.V.; Dehmlow, S.S.; Phase Transfer Catalysis, Verlag Chemie, Weinheim, 1980.
4. Starks, C.M.; Napier, D.R.; British Patent 1,227,144 filed April 5, 1967.
5. Makosza, M.; Serafinova, B.; Rocz. Chem. 1965, 39, 1223.
6. Starks, C. M.; J. Am. Chem. Soc. 1971, 93, 195.
7. Makosza, M.; Wawrzyniewicz, M.; Tetrahedron Lett. 1969, 4659.
8. Starks, C.M.; Owens, R.M.; J. Am. Chem. Soc. 1973, 95, 3613.
9. Aldrich Chemical Co., 1986-87 catalog.
10. Gokel, G.W.; Cram, D.J.; Liotta, C.L.; Harris, H.P.; Cook, F.L.; J. Org. Chem. 1974, 39, 2445.
11. Pedersen, C.J.; Org. Synth. 1972, 52, 66.
12. Leong, B.K.J.; Ts'o, T.O.T.; Chenoweth, M.B.; Toxicol. Appl. Phamacol. 1974 27, 342.
13. Takayama, K.; Hasegawa, S.; Sasegawa, S.; Namber, N.; Ngahai, T.; Chem. Pharm. Bull. 1977, 25, 3125.
14. Hendrixson, R.R.; Mack, M.P.; Palmer, R.A.; Ottolenghi, A.; Ghirardelli, R.G.; Toxicol. Appl. Pharmacol. 1978, 44, 263.
15. Kato, N.; Ikeda, I.; Okahara, M.; Shibasaki, I.; J. Antibact. Antifung. Agents 1980, 8, 415, 532.
16. Achenback, C.; Hauswirth, O.; Kossmann, J.; Ziskoven, R.; Physiol. Chem. Phys. 1980, 12, 277.
17. Tso, W.-W.; Fung, W.-P.; Tso, M.-Y.; J. Inorg. Biochem. 1981, 14, 237.
18. Tso, W.-W.; Fung, W.-P.; Inorg. Chim. Acta 1981, 55, 129.

19. Tso, W.-W.; Fung, W.-P.; <u>Microbios. Lett.</u> **1982** <u>19</u>, 145.
20. Kolbeck, R.C.; Bransome, E.D.; Spier, W.A.; Hendry, L.B.; <u>Experentia</u> **1984**; <u>40</u>, 727.
21. Dean, B.J.; Brooks, T.M.; Hodson-Walker, G.; Hutson, D.H.; <u>Mutation Res.</u> **1985**, <u>153</u>, 57.
22. Gokel, G. W.; Goli, D. M.; Minganti, C.; Echegoyen, L.; <u>J. Am. Chem. Soc.</u> **1983**, <u>105</u>, 6786.
23. Wipff, G.; Weiner, P.; Kollman, P; <u>J. Am. Chem. Soc.</u> **1982**, <u>104</u>, 3249.
24. Michaux, G.; Reisse, J.; <u>J. Am. Chem. Soc.</u> **1982**, <u>104</u>, 6895.
25. Perrin, R.; Decoret, C.; Bertholon, G.; Lamartine, R.; <u>Nouveau J. Chim.</u> **1983**, <u>7</u>, 263.
26. Izatt, R.M.; Bradshaw, J.S.; Nielsen, S.A.; Lamb, J.D.; Christensen, J.J.; Sen, D.; <u>Chem. Rev.</u> **1985**, <u>85</u>, 271.
27. Gokel, G. W.; Goli, D. M.; Schultz, R. A.; <u>J. Org. Chem.</u> **1983**, <u>48</u>, 2837 and references therein.
28. Harris, J.M.; Hundley, N.H.; Shannon, T.G.; Struck, E.C.; <u>J. Org. Chem.</u> **1982**, <u>47</u>, 4789.
29. Gokel, G. W.; Dishong, D. M.; Diamond, C. J.; <u>J. Chem. Soc. Chem. Commun.</u> **1980**, 1053.
30. Dishong, D.M.; Diamond, C.J.; Cinoman, M.I.; Gokel, G.W.; <u>J. Am. Chem. Soc.</u> **1983**, <u>105</u>, 586.
31. Schultz, R. A.; White, B.A.; Dishong, D. M.; Arnold, K.A.; Gokel, G. W.; <u>J. Am. Chem. Soc.</u> **1985**, <u>107</u>, 6659.
32. Furukawa, N.; Imaoka, K.; Fujihara, H.; Oae, S.; <u>Chemistry Lett.</u> **1982**, 1421.
33. Kondo, S.; Ohta, K.; Tsuda, K.; <u>Makromol. Chem. Rapid Commun.</u> **1983**, <u>4</u>, 145.
34. Kondo, S.; Ohta, K.; Ojika, R.; Yasui, H.; Tsuda, K.; <u>Makromol. Chem.</u> **1985**, <u>186</u> 1.
35. Tomoi, M.; Ogawa, E.; Hosokawa, Y.; Kakiuchi, H.; <u>J. Polym. Sci.: Polym. Chem. Ed.</u> **1982** <u>20</u>, 3015.
36. Tomoi, M.; Ogawa, E.; Hosokama, Y.; Kakiuchi, H.; <u>J. Polym. Sci.: Polym. Chem. Ed.</u> **1982**, <u>20</u>, 3421.
37. Tomoi, M.; Hosokawa, Y.; Kakiuchi, H.; <u>Makromol. Chem., Rapid Commun.</u> **1983**, <u>4</u>, 227.
38. Frechet, J.M.J.; Kelly, J.; Sherrington, D.C.; <u>Polymer</u> **1984** <u>25</u>, 1491.
39. Kelly, J.; Sherrington, D.C.; <u>Polymer</u> **1984** <u>25</u>, 1499.
40. Ford, W.T.; Periyasamy, M.; Spivey, H.O.; <u>Macromolecules</u> **1984**, <u>17</u>, 2881.
41. Tomoi, M.; Nakamura, E.; Hosokawa, Y.; Kakiuchi, H.; <u>J. Polym. Sci.: Polym. Chem. Ed.</u> **1985**, <u>23</u>, 49.
42. Hradil, J.; Svec, F.; <u>Polymer Bulletin</u> **1983**, <u>10</u>, 14.
43. Heffernan, J.G.; Sherrington, D.C.; <u>Tetrahedron Lett.</u> **1983** <u>24</u>, 1661.
44. Kimura, Y.; Regen, S.L.; <u>J. Org. Chem.</u> **1983**, <u>48</u>, 195.
45. Kimura, Y.; Kirszensztejn, P.; Regen, S.L.; <u>J. Org. Chem.</u> **1983**, <u>48</u>, 385.
46. Arkles, B.; Peterson, Jr., W.B.; King, K.; <u>A.C.S. Symposium Series,</u> Vol. <u>192</u> **1982**, 281.
47. Anzai, J.; Sakata, Y.; Ueno, A.; Osa, T.; <u>Makromol. Chem. Rapid Commun.</u> **1982**, <u>3</u>, 399.
48. Blasius, E.; Janzen, K.-P.; Klotz, H.; Toussaint, A.; <u>Makromol. Chem.</u> **1982**, <u>183</u>, 1401.

48. Pannell, K.H.; Mayr, A.J.; J. Chem. Soc. Perkin Trans. I **1982**, 2153.
49. Tomoi, M.; Yanai, N.; Shiiki, S.; Kakiuchi, H.; J. Polym. Sci.: Polym. Chem. Ed. **1984**, 22, 911.
50. Anelli, P.L.; Czech, B.; Montanari, F.; Quici, S.; J. Am. Chem. Soc. **1984**, 106, 861.
51. Cornelis, A.; Laszlo, P.; Synthesis **1982**, 162.
52. Tundo, P.; Venturello, P.; Angeletti, J. Am. Chem. Soc. **1982**, 104, 6551.
53. Sawicki, R.; Tetrahedron Lett. **1982**, 2249.
54. Kaifer, A.; Echegoyen, L.; Durst, H.; Schultz, R.A.; Dishong, D.M.; Goli, D.M.; Gokel, G.W.; J. Am. Chem. Soc. **1984**, 106, 5100.
55. Kaifer, A.; Echegoyen, L.; Gokel, G.W.; J. Org. Chem. **1984**, 49, 3029.
56. Schultz, R.A.; Schlegel, E.; Dishong, D.M.; Gokel, G.W.; J. Chem. Soc. Chem. Commun. **1982**, 242.
57. Fronczek, F.R.; Gatto, V.J.; Schultz, R.A.; Jungk, S.J.; Colucci, W.J.; Gandour, R.D.; Gokel, G.W.; J. Am. Chem. Soc. **1983**, 105, 6717.
58. Fronczek, F.R.; Gatto, V.J.; Minganti, C.; Schultz, R.A.; Gandour, R.D.; Gokel, G.W.; J. Am. Chem. Soc. **1984**, 106, 7244.
59. Gandour, R.D.; Fronczek, F.R.; Gatto, V.J.; Minganti, C.; Schultz, R.A.; White, B.D.; Arnold, K.A.; Mazocchi, D.; Miller, S.R.; Gokel, G.W.; J. Am. Chem. Soc. **1986**, 108, 0000.
60. Gatto, V. J.; Gokel, G. W.; J. Am. Chem. Soc. **1984**, 106, 8240.
61. Gatto, V.J.; Arnold, K.A.; Viscariello, A.M.; Miller, S.R.; Gokel, G.W.; Tetrahedron Lett. **1986**, 327.

RECEIVED July 26, 1986

Chapter 5

Stable Catalysts for Phase Transfer at Elevated Temperatures

Daniel J. Brunelle

General Electric Corporate Research and Development, Schenectady, NY 12301

> The utilization of polar polymers and novel N-alkyl-4-(N',N'-dialklamino)pyridinium salts as stable phase transfer catalysts for nucleophilic aromatic substitution are reported. Polar polymers such as poly(ethylene glycol) or polyvinylpyrrolidone are thermally stable, but provide only slow rates. The dialkylaminopyridininium salts are very active catalysts, and are up to 100 times more stable than tetrabutylammonium bromide, allowing recovery and reuse of catalyst. The utilization of bis-dialkylaminopyridinium salts for phase-transfer catalyzed nucleophilic substitution by bisphenoxides leads to enhanced rates, and the requirement of less catalyst. Experimental details are provided.

Interests in phase transfer catalysis (PTC) have grown steadily for the past several years (1-3). The technique has achieved industrial importance in cases where the alternative use of polar aprotic solvents would be prohibitively expensive (4-7). Although substantial efforts in the synthesis or implementation of new catalysts have been made in the areas of crown ether chemistry (8,9 also see 10-12) and polymer-supported catalysts (13-16), very few examples of catalysts which extend the realm of PTC chemistry into the regime of high temperature reactions (110-200°) have been identified. Our interests in nucleophilic aromatic substitution reactions have stimulated research in this area. Crown ethers are stable PTC agents, bulk industrial use of which is precluded due to their high cost and possible toxicity (10). Phosphonium salts are sometimes more stable than ammonium salts, but are also susceptible to S_N2 attack by strong nucleophiles (17). We have investigated several categories of phase transfer catalysts, with the objective of identifying PTC agents which are stable to the elevated temperatures and strongly nucleophilic conditions encountered during nucleophilic aromatic substitution.

0097-6156/87/0326-0038$06.00/0
© 1987 American Chemical Society

Discussion

We have been interested in nucleophilic aromatic displacement reactions by phenoxide or thiophenoxide, reactions with potential applications to monomer synthesis, for several years (Equation 1). Although phase transfer catalyzed nucleophilic aromatic substitution reactions are known for very activated systems [such as the reaction of 4-nitro-N-methylphthalimide with phenoxides, (5) or substitutions with carbanions in doubly activated aromatic systems (18-20)], reactions on less activated systems have proven unsuccessful. For example, phase transfer catalyzed displacement of 4-chloronitrobenzene by phenoxide has been shown to afford low-yields, unless a full equivalent of tetrabutylammonium bromide is used (21). When we studied this reaction, we found that Bu_4NBr was quickly consumed by both Hoffman elimination and S_N2 displacement reactions, under the conditions of the reaction. Use of catalytic amounts of Bu_4NBr (5%) led to only about 12% of the desired product, even though the initial reaction rate was high. Although we found crown ethers to be acceptable catalysts, limitations due to their cost and toxicity forced us to turn elsewhere. Polar polymers such as poly(ethylene glycol) [PEG], PEG ethers and polyvinylpyrrolidone [PVP] have been described as PTC reagents for a variety of reactions, including displacement of alkyl halides by phenoxide, alkylation of aliphatic alkoxides, and attack of thiolates on alkyl and aryl halides (22-26). However, in many of these cases, use of alternative PTC agents such as quaternary ammonium salts results in more efficient reactions. For example, displacement of butyl bromide by potassium phenoxide requires reflux in toluene using PEG as the PTC (22), whereas the reaction proceeds to completion in 2-12 hr at ambient temperature using benzyltributylammonium bromide as the catalyst (27). We report here, however, that polar polymers, such as PEG, crosslinked PVP, and especially poly(ethylene glycol methyl ether) [PEGM], can be efficient catalysts for nucleophilic aromatic substitution reactions.

$$\underset{\substack{X = F, Cl \\ Y = NO_2, CN, SO_2Ph, COPh}}{\ce{X-C_6H_4-Y}} \xrightarrow[100-200°]{Nu^-, PTC} \underset{Nu = ArO^-, ArS^-, S^{-2}}{\ce{Nu-C_6H_4-Y}} \quad (1)$$

Results

Figure 1 shows a plot of yield vs time for the reaction of 4-chloronitrobenzene with potassium cresoxide, catalyzed by Bu_4NBr, PEGM 5000 (polyethylene glycol monomethyl ether, avg. molecular weight = 5000) or PVP-X (crosslinked polyvinylpyrrolidone). Although the reaction catalyzed by the ammonium salt has a high

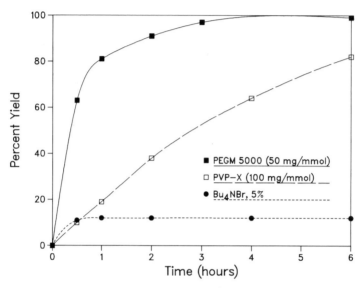

Figure 1. Reaction forming 4-cresyl-4'-nitrophenyl ether from 4-chloronitrobenzene and potassium cresoxide in chlorobenzene at 125° catalyzed by PEGM 5000, PVP-X, or Bu_4NBr.

initial rate, decomposition of the catalyst (via S_N2 reaction with cresoxide) occurs within a short time, and the yield does not increase after the initial hour of reaction. Reaction using PEGM proceeds smoothly with no catalyst deactivation, affording high yields of 4-cresyl-4'-nitrophenyl ether after several hours. Reactions catalyzed by PVP-X are 2-3 times slower, but also afford high yields of aryl ether; reaction at 150° affords a 92% yield of nitrophenyl ether in 4 hr.

Potassium phenoxides react 2-4 times faster than the sodium salts using these catalysts. We have attempted to enhance the rate of displacement by adding small amounts of potassium salt to reactions utilizing sodium phenoxides. Curiously, the opposite effect occurs: small amounts of sodium cresoxide appear to poison the catalyst; the rate of reaction is identical to that obtained using pure sodium cresoxide, even when a 75:25 mixture of potassium/sodium cresoxide is used. Apparently, the sodium ion is preferentially bound to the polymer matrix, and provides a less reactive form of phenoxide (For details of binding of oligoethylene glycols with sodium cations, see 28.). Only when sodium is completely excluded from the reaction medium is an advantageous rate enhancement seen (Table I).

Table I. Effect of Cation on PVP
Catalyzed Reaction of Sodium or Potassium
Cresoxide with 4-Chloronitrobenzene[a]

%K	%Na	Yield (1hr)[b]
0	100	23
100	0	43
50	50	22
67	33	23
75	25	24
90	10	28

[a] Reaction in o-dichlorobenzene at 125°.
[b] Yield by VPC analysis after reaction for 1 hr.

Further evidence for the stability of these polar polymers to severe conditions was obtained from extended reactions at high temperatures. For example, reaction of anhydrous sodium sulfide with chlorophenyl phenyl sulfone in o-dichlorobenzene at 180° using PEGM 5000 as the catalyst affords a 74% yield of bis-sulfone sulfide after 72 hr. Other examples using PEGM or PVP as catalysts for displacement with phenoxides or thiophenoxides are detailed in Table II. Yields are generally very high for reactions catalyzed by 25-50 mg of PEGM 5000 per mmol of reactant. Recovery of the catalyst can be achieved by precipitation with diethyl ether (29).

Dialkylaminopyridinium Salts. Although phase transfer catalysis by polar polymers provided displacement reactions under conditions where conventional catalysts such as Bu_4NBr decompose, rates of reaction, a preference for potassium over sodium salt reactants, and amounts of catalyst necessary for convenient reaction were

Table II. Phase Transfer Catalyzed Nucleophilic Aromatic Displacement Reactions[a]

Substrate	Nucleophile [b]	Catalyst (amt)[c]	Time (hr) / temp	Product (Yield) [d]
Cl–C₆H₄–NO₂	NaOCr NaOCr NaOCr	PEGM 5000 (25) Bu₄NBr (5) PVP-X (100)	0.5/ 125 16/ 125 8/ 125	PhO–C₆H₄–NO₂ (97) (15) (95)
Cl–C₆H₄–NO₂	NaSCr	PEGM 5000 (25)	1/ 125	CrS–C₆H₄–NO₂ (98)
4-Cl-C₆H₄–C(O)–C₆H₅	NaOCr	PEGM 5000 (50)	1/ 125	4-CrO-C₆H₄–C(O)–C₆H₅ (98)
4-Cl-C₆H₄–SO₂–C₆H₅	NaSCr NaSCr	PEGM 5000 (25) Bu₄NBr (10)	2/ 125 12/ 125	4-CrS-C₆H₄–SO₂–C₆H₅ (98) (3)
4-Cl-C₆H₄–SO₂–C₆H₅	Na₂S	PEGM 5000 (50)	72/ 180[e]	(C₆H₄–SO₂–C₆H₅)₂S (60)
F–C₆H₄–CN	NaSCr NaSCr	PEGM 5000 (50) Bu₄NBr (10)	1/ 150[e] 16/ 150[e]	CrS–C₆H₄–CN (92) (5)

[a]Reactions were carried out in chlorobenzene, except where noted. [b]Cr = 4-Cresyl. [c]Catalyst level of polymers given in mg/mmol of substrate; Bu₄NBr amounts are equivalents. [d]Isolated yields. [e]o-Dichlorobenzene as solvent.

serious limitations. Thus we turned our attention to preparation of new ammonium salts which might be more stable than tetra-n-alkyl ammonium halides.

We first attempted to make the ammonium salts more stable using steric hindrance. We prepared a variety of sterically encumbered ammonium salts via the sequence shown in Equation 2. Although we could prepare a variety of hindered tertiary amines, successful quaternizations were achieved only with methyl groups. These salts were more stable to phenoxide than n-Bu$_4$NBr by factors of 10-50, demonstrating that steric hindrance can play a significant part in stabilization of phase transfer catalysts. However, none of these salts were more effective than n-Bu$_4$NBr as phase transfer agents, probably due to their less lipophilic nature, and we turned our attention to other systems.

$$(2)$$

We next became interested in the possibility of making ammonium salts more chemically resistant via resonance stabilization. We suspected that salts of 4-dialkylaminopyridines might be more stable than typical ammonium salts, since the positive charge is shared among several atoms, rather than being concentrated on a single nitrogen atom. In addition, we anticipated that building steric hindrance into salts of this type would further add to their stability in the presence of strongly nucleophilic reactants. Thus, we prepared a series of N-alkyl-N',N'-dialkylaminopyridinium salts (1-5), and tested both their stability and PTC efficiency.

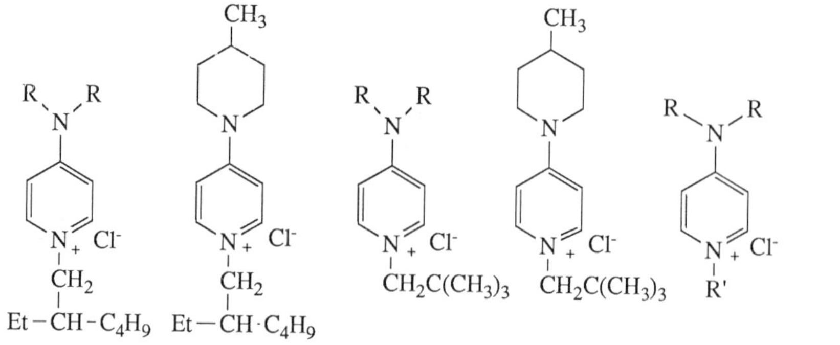

Structures

1
R₂N–(pyridine)–N⁺(CH₂–CH(Et)–C₄H₉) Cl⁻
- a: R = Me
- b: R = n-Bu
- c: R = n-Hexyl

2
(4-methylpiperidinyl)–(pyridine)–N⁺(CH₂–CH(Et)–C₄H₉) Cl⁻

3
R₂N–(pyridine)–N⁺(CH₂C(CH₃)₃) Cl⁻
- a: R = Me
- b: R = n-Bu
- c: R = n-Hexyl

4
(4-methylpiperidinyl)–(pyridine)–N⁺(CH₂C(CH₃)₃) Cl⁻

5
R₂N–(pyridine)–N⁺(R') Cl⁻
- a: R = Me; R' = n-Bu
- b: R = n-Bu; R' = n-Bu
- c: R = Me; R' = n-octyl

Alkylation of p-dialkylaminopyridines [4-dimetehylamino pyridine and 4-(4-methyl-piperidinyl)-pyridine are commercially available; 4-dibutyl and 4-dihexylaminopyridines were prepared from the dialkylamine, phosphorus pentoxide, and 4-hydroxypyridine, according to ref. 30] with n-alkyl halides proceeds within 1 hr in toluene at 50°. Alkylation with 2-ethylhexyl mesylate requires 2 hr of reflux in toluene, and alkylation with neopentyl mesylate requires reaction for 72 hr neat at 130°. Metathesis to the chloride or bromide may be carried out via ion exchange or by washing methylene chloride solutions of the mesylate salts with saturated NaCl or NaBr. Table III summarizes the data obtained on the stability of various catalysts in reactions of equimolar amounts of catalyst and sodium phenoxide in toluene or chlorobenzene. To our delight, the N-alkyl-p-dialkylaminopyridinium salts were significantly more stable than Bu₄NBr.

Table III. Catalyst Half-lives in Presence of Sodium Phenoxide

Catalyst	Solvent	Temp	Half-life
Bu₄NBr	toluene	110°	7 min
1a	toluene	110°	8 hr
2	toluene	110°	11 hr
2	PhCl	125°	9 hr
3a	toluene	110°	12 hr
5c	toluene	110°	2 hr

We next examined the PTC efficiency of these catalysts, and found that these dialkylaminopyridinium salts were superior catalysts. Nucleophilic aromatic displacement reactions using these catalysts are summarized in Table IV. These catalysts are effective even at temperatures as high as 180°, in displacements utilizing phenoxide or thiophenoxide nucleophiles. Comparison of several PTC agents for the displacement of 4-chloronitrobenzene with sodium phenoxide is made in Figure 2, for reactions carried out at 125° in chlorobenzene. Although 18-crown-6, Bu_4PBr, or PEGM 5000 are also reasonably stable to displacement conditions, they are much less efficient catalysts on a molar basis.

Unfortunately, these dialkylaminopyridinium salts have limited utility in the presence of aqueous NaOH or Na_2S, even at moderate temperatures. Displacement of dialkylamine and formation of 2-alkylpyridone or thiopyridone occurs within hours at 100° (Equation 3).

$$\underset{\underset{\underset{CH_3CH_2}{|}}{\overset{|}{CH_2-CH-C_4H_9}}}{\overset{\overset{R\diagdown N\diagup R}{|}}{\underset{N^+\ Cl^-}{\bigcirc}}} \xrightarrow[H_2O]{NaOH\ or\ Na_2S} \underset{\underset{\underset{CH_3CH_2}{|}}{\overset{|}{CH_2-CH-C_4H_9}}}{\overset{\overset{Y}{||}}{\underset{N}{\bigcirc}}} + HNR_2 \quad (3)$$

Y = O, S

One of the goals in developing stable PTC reagents was to minimize the amount of catalyst necessary for complete reaction. Even though the dialkylaminopyridinium salts were substantially more stable than Bu_4NBr, reactions became inconveniently slow at low catalyst levels, when utilizing dianionic substrates, such as the disodium salt of bisphenol A. It appeared that such reactions of dianionic species were second order in catalyst; i.e. one ammonium salt per phenoxide is necessary for phase transfer. Thus, decreasing the catalyst level by a factor of five leads to a rate decrease by a factor of 25. Because of our interest in reactions of dianions, we have now examined dialkylaminopyridinium bis-salts, anticipating that choice of the proper chain length connecting the salt tails might lead to reactions unimolecular in catalyst (Figure 3). Several catalysts of this type were prepared and tested.

Table IV. Phase Transfer Catalyzed Nucleophilic Aromatic Displacement Reactions[a]

Substrate	Nucleophile	Catalyst (%)[b]	Time/temp	Product (yield)[c]
F–C$_6$H$_4$–NO$_2$	PhONa	2 (0.5)	15 min/110[d]	PhO–C$_6$H$_4$–NO$_2$ (97)
Cl–C$_6$H$_4$–NO$_2$	PhONa	3a (5)	30 min/125[e]	PhO–C$_6$H$_4$–NO$_2$ (95)
Cl–C$_6$H$_4$–NO$_2$	PhONa	Bu$_4$NBr (10)	8hr/125[e]	PhO–C$_6$H$_4$–NO$_2$ (12)
Cl–C$_6$H$_4$–CN	PhSNa	1a (5)	4hr/150[f]	PhS–C$_6$H$_4$–CN (86)
Cl–C$_6$H$_4$–SO$_2$–C$_6$H$_4$–Cl	PhONa	2 (5)	2hr/180[f]	PhO–C$_6$H$_4$–SO$_2$–C$_6$H$_4$–OPh (87)
Cl–C$_6$H$_4$–SO$_2$–C$_6$H$_4$–Cl	PhSNa	1a (5)	2 hr/150[f]	PhS–C$_6$H$_4$–SO$_2$–C$_6$H$_4$–SPh (90)
Cl–C$_6$H$_4$–CO–C$_6$H$_4$–Cl	PhSNa	1b (5)	2 hr/110[d]	PhS–C$_6$H$_4$–CO–C$_6$H$_4$–SPh (90)

[a]All reactions were carried out under N$_2$, using anhydrous conditions, with 5% nucleophile. [b]Mole % relative to substrate. [c]Isolated Yields. [d]Toluene solvent. [e]Chlorobenzene solvent. [f]o-Dichlorobenzene solvent.

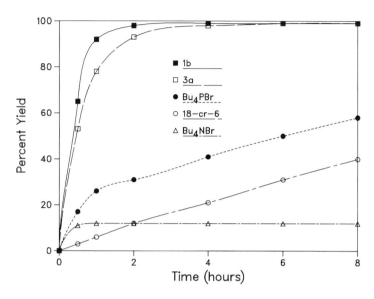

Figure 2. Preparation of 4-nitrophenyl phenyl ether from sodium phenoxide and 4-chloronitrobenzene using 5 mole % catalyst in chlorobenzene at reflux.

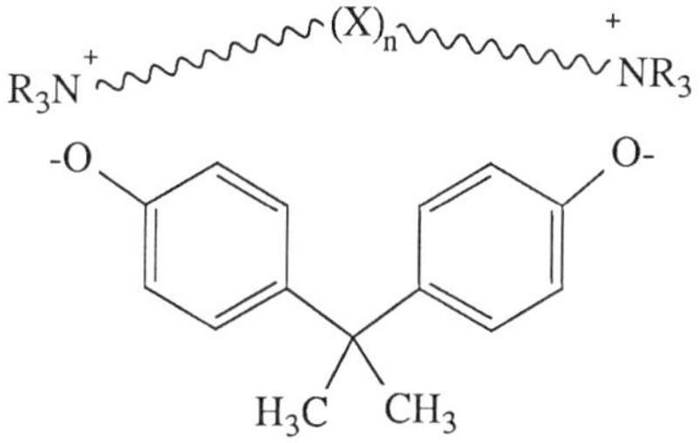

Figure 3. Bisammonium salt phase-transfer catalyst.

The data plotted in Figure 4 dramatically exemplifies the effect of reducing catalyst levels for a pyridinium salt vs a bispyridinium salt. It is clearly apparent that the rate of reaction does not drop nearly as rapidly for the bis-salt when the catalyst level is decreased. Although exact first order kinetics are not observed, the reaction rate appears to be nearer first than second order in catalyst. It is also apparent that the bis-salts are more than twice as effective as the mono-salts on a molar basis. Results with displacements of other dianions or substrates are summarized in Table V.

Conclusion

Spurred by our desire to avoid use of expensive dipolar aprotic solvents in nucleophilic aromatic substitution reactions, we have developed two alternative phase transfer systems, which operate in non-polar solvents such as toluene, chlorobenzene, or dichlorobenzene. Polar polymers such as PEG are inexpensive and stable, albeit somewhat inefficient PTC agents for these reactions. N-Alkyl-N',N'-Dialkylaminopyridinium salts have been identified as very efficient PTC agents, which are about 100 times more stable to nucleophiles than Bu_4NBr. The bis-pyridinium salts of this family of catalysts are extremely effective for phase transfer of dianions such as bis-phenolates.

Experimental Section

General. Toluene, chlorobenzene, and o-dichlorobenzene were distilled from calcium hydride prior to use. 4-Dimethylaminopyridine (Aldrich Chemical Co) was recrystalled (EtOAc), and the other 4-dialkylaminopyridines were distilled prior to use. PEG's, PEGM's, PVP's, and crown ethers were obtained from Aldrich Chemical Co., and were used without purification. Bu_4NBr and Bu_4PBr were recrystallized (toluene). A Varian 3700 VPC interfaced with a Spectraphysics SP-4000 data system was used for VPC analyses. A Dupont Instruments Model 850 HPLC (also interfaced with the SP-4000) was used for LC analyses. All products of nucleophilic aromatic substitution were identified by comparison to authentic material prepared from reaction in DMF or DMAc. Alkali phenolates or thiolates were pre-formed via reaction of aqueous NaOH or KOH and the requisite phenol or thiophenol in water under nitrogen, followed by azeotropic removal of water with toluene. The salts were transferred to jars under nitrogen, and were dried at $120°$ under vacuum for 20 hr, and were stored and handled in a nitrogen dry box.

Phase Transfer Catalysts. p-Dialkylaminopyridines were alkylated in toluene by stirring equimolar quantities of pyridine and alkyl halide or mesylate. Alkylation with n-alkylbromides required 1 hr at $50°$. Alkylation with 2-ethylhexyl mesylate required reflux in toluene for 2 hr. Neopentylation using neopentyl mesylate required heating the reagents neat at $130°$ for 72-100 hr. The neopentyl salts required the following workup: 1) basification with 2 M NaOH, extracting the salt into the aqueous phase, and washing unreacted starting materials away with petroleum ether; 2)

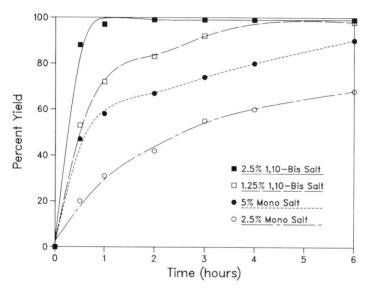

Figure 4. Preparation of the bis-(4-nitrophenyl) ether of bisphenol A via reaction of bisphenol A disodium salt with 4-chloronitrobenzene. Bis salt = bis-(4-dihexylaminopyridinium) decane dibromide. Mono salt = N-(2-ethylhexyl)- 4-dihexylaminopyridinium mesylate.

Table V. Aryl Displacement using 1,10-bis-(4-Dihexylaminopyridinium) Decane Dibromide as PTC[a]

Starting Materials	Product	Yield
NaO–C6H4–C(CH3)2–C6H4–ONa + Cl–C6H4–NO2	O2N–C6H4–O–C6H4–C(CH3)2–C6H4–O–C6H4–NO2	99%
1,3-(ONa)2–C6H4 + Cl–C6H4–NO2	O2N–C6H4–O–C6H4(1,3)–O–C6H4–NO2	97%
NaO–C6H4–S–C6H4–ONa + Cl–C6H4–NO2	O2N–C6H4–O–C6H4–S–C6H4–O–C6H4–NO2	99%
NaO–C6H4–C(CH3)2–C6H4–ONa + Cl–C6H4–C(=O)–C6H4–Cl	(–C6H4–O–C6H4–C(=O)–C6H4–O–C6H4–C(CH3)2–)n, n=10	87%

[a]Reactions were carried out in chlorobenzene at reflux, using 2.5% (molar) catalyst relative to bisphenolate.

acidification with conc. HCl or HBr, and extraction of the salt into CH_2Cl_2; 3) recrystallization from $CH_3CN/EtOAc$. All other salts (except the dihexylaminopyridinium salts, which were oils) were recrystallized from $CH_3CN/EtOAc$, and were dried at 120° under vacuum prior to use. The salts are fairly hygroscopic, and should be kept in a dry box. Specific examples:

N-(2-ethylhexyl)-4-Dimethylaminopyridinium mesylate.
2-Ethylhexyl mesylate (20.8 g; 100 mmol) was heated at reflux under nitrogen with 12.22 g (100 mmol) of 4-dimethylaminopyridine in 100 ml of toluene solution for 3 hr. Toluene was removed by a rotary evaporator leaving the crude product, a somewhat hygroscopic light tan semisolid.

Chloride salt (1a).
The pyridinium mesylate from above was dissolved in 200 ml of methylene chloride, and was shaken with 50 ml of saturated NaCl solution in a separatory funnel for ca 2 min. The methylene chloride phase was removed, the brine phase was washed once with 20 ml of methylene chloride, and the organic phases were combined. This procedure was repeated twice using fresh brine. The methylene chloride phase was filtered and evaporated to dryness, affording the crude chloride salt. Recrystallization of the salt from THF/chloroform (ca. 20:1) gives white plates with mp = 192-193°.

(N-Neopentyl)-4-Dihexylaminopyridinium Bromide (3b).
The neopentyl salt was prepared in a similar manner from neopentyl mesylate, but reaction was carried out neat at 130° for 72 hr. Higher temperatures cannot be used, due to decomposition of neopentyl mesylate. The crude product was dissolved in water, basified to neutralize any pyridinium salt, and was washed with petroleum ether to remove amine and unreacted neopentyl mesylate. The aqueous phase was acidified with HBr, and extracted with methylene chloride, to afford crude salt. Recrystallization from 20:1 ethyl acetate/acetonitrile affords the product (mp = 169-170°).

Bis-(4-dihexylaminopyridinium)decane dibromide.
4-Dihexylaminopyridine (2.63 g; 10.0 mmol) and 1,10-dibromodecane were refluxed in CH_3CN for 2 hr. Acetonitrile was removed by distillation and replaced with toluene. The product is an oil.

General Procedure for PTC Catalyzed Aryl Displacement Reactions.
Alkali phenolate or thiolate, substrate, phase transfer catalyst, and a magnetic stirring bar were weighed into a dry stoppered flask under nitrogen. Solvent was added, and the reaction mixture was stirred and heated under the conditions described in the tables. Hydrocarbon internal standards (n-alkanes for VPC analysis, anthracene or m-terphenyl for HPLC analysis) were used for yield determinations. Reactions (or reaction aliquots) were quenched with HOAc, diluted with CH_2Cl_2, and filtered through short pads of silica gel. Recovery of dialkylaminopyridinium salts could be achieved by washing the toluene, chlorobenzene or o-dichlorobenzene solutions with water (partition into water >10:1 from these solvents), followed by extraction from water into CH_2Cl_2 (partition into CH_2Cl_2 from water >5:1).

Acknowledgments I thank Daniel A. Singleton for technical assistance, and Paul Donahue and Steven Dorn for NMR and MS spectral analyses. I also thank Reilly Tar and Chemical Co. for a generous sample of 4-(4-methylpiperidinyl)pyridine.

Literature Cited

1. Weber, W.P.; Gokel, G.W. "Phase Transfer Catalysts in Organic Synthesis," Springer-Verlag: W. Berlin, 1977.
2. Starks, C.M.; Liotta, C.L. "Phase Transfer Catalysis," Academic Press: New York, New York, 1978.
3. Dehmlow, E.V.; Dehmlow, S.S. "Phase Transfer Catalysis," Verlag Chimie: W. Berlin, 1980.
4. Haloalkyl alkyl ethers: Alt, G.H.; Chepp, J.P. U.S. Patent 4 371 717, 1978.
5. Armomatic etherimides: Williams, F.J. U.S. Patent 4 273 712, 1981.
6. Cyclic polyformals: Johnson, D.S. U.S. Patent 4 163 833, 1979.
7. Aromatic polyesters: Rieder, W. U.S. Patent 4 430 493, 1984.
8. Hiraoka, M. "Crown Compounds, their characteristics and applications," Elsevier Sci. New York, New York, 1982.
9. Gokel, G.W.; Korzeniowski, S.H. "Macrocyclic Polyether Synthesis," Springer-Verlag, New York, New York, 1982.
10. Dishong, D.M.; Diamond, C.J.; Cinoman, M.I.; Gokel, G.W. $\underline{J}$. $\underline{Am}$. $\underline{Chem}$. $\underline{Soc}$., 1983, 105, 587.
11. Bogatskii, A.V.; Luk, Y. N. G.; and Pastushok, V.N.$\underline{Dokl}$. $\underline{Akad}$. $\underline{Nauk}$ $\underline{SSSR}$, 1983, 271, 1392.
12. Inokuma, S.; Kuwamura, T.; $\underline{Nippon}$ $\underline{Kagaku}$ $\underline{Kaishi}$, 1983, 1494.
13. Regen, S.L.; Kimura, Y. $\underline{J}$. $\underline{Am}$. $\underline{Chem}$. $\underline{Soc}$., 1982, 104, 2064.
14. Mathur, N.K.; Narang, C.K.; Williams, R.E. "Polymers as Aids in Organic Chemistry," pp 19, 209-213, Academic Press, New York, New York, 1980.
15. Reference 3, pp 45, 55-57.
16. Hodge, P.; Sherrington, D.C.; "Polymer-supported Reactions in Organic Synthesis," John Wiley, New York, New York, 1980.
17. Brunelle, D.J. $\underline{Chemosphere}$, 1983, 12, 167.
18. Makosza, M.; Jagusztyn-Grochowska, M.; Ludwikow, M.; Jawdosiuk, M. $\underline{Tetrahedron}$, 1974, 30, 3723, and references therein.
19. Jawdosiuk, M.; Makosza, M.; Malinowski, E.; Wilczynski, W. $\underline{Pol}$. $\underline{J}$. $\underline{Chem}$., 1978, 52, 2189, 2189.
20. Frimer, A.; Rosenthal, I. $\underline{Tetrahedron}$ $\underline{Lett}$., 1976, 2809.
21. Paradisi, C.; Quintily, U.; Scorrano, G. $\underline{J}$. $\underline{Org}$. $\underline{Chem}$., 1983, 48, 3022.
22. Kelly, J.; MacKenzie, W.M.; Sherrington, D.C. $\underline{Polymer}$, 1979, 20, 1048.
23. Vogtle, F.; Weber, E. $\underline{Angew}$. $\underline{Chemie}$ $\underline{Int}$. $\underline{Ed}$. $\underline{Eng}$., 1979, 18, 753
24. Lehmkuhl, H.; Rabet, F.; Hauschild, K. $\underline{Synthesis}$, 1977, 1984.
25. Balasubramanian, D.; Sukumar, P.; Chandani, B. $\underline{Tetrahedron}$ $\underline{Lett}$., 1979, 3543.
26. Neumann, R.; Sasson, Y. $\underline{Tetrahedron}$, 1983, 39, 3437.
27. McKillop, A.; Fiaud, J.C.; Hug, R.P. $\underline{Tetrahedon}$, 1974, 30, 1379.
28. Gokel, G.W.: Goli, D.M.; Schultz, R.A. $\underline{J}$. $\underline{Org}$. $\underline{Chem}$., 1983 48, 2837.

29. Harris, J.M.; Hundley, N.H.; Shannon, T.G. J. Org. Chem., 1982, 47, 4789.
30. Pederson, E.B.; Carlson, D. Synthesis, 1978, 844.

RECEIVED July 26, 1986

Chapter 6

Reactivity and Application of Soluble and Polymer-Supported Phase-Transfer Catalysts

Fernando Montanari, Dario Landini, Angelamaria Maia, Silvio Quici, and Pier Lucio Anelli

Centro C.N.R. and Dipartimento di Chimica Organica e Industriale dell'Università, Milano, Italy

A survey of molecules capable of working as phase-transfer catalysts is presented. Special attention is devoted to quaternary onium salts, crown-ethers and cryptands, all of which have a high lipophilicity ensured by aliphatic chains. An examination is made of the structural factors and reaction conditions ruling the reactivity of anions associated with these systems. Concentrated aqueous solutions of alkali hydroxides affect the hydration and hence the reactivity of anions transferred by lipophilic cations into the organic phase, an effect strongly enhanced in the case of OH^-. Lipophilic cage ligands forming very stable complexes with sodium salts are particularly efficient catalysts for reactions promoted by highly hydrophilic and slightly polarizable anions, like CH_3O^-, OH^-, F^-, etc. Polymer-supported phase-transfer catalysts (quaternary salts, crown-ethers and cryptands) are examined and their reactivity is compared with that of the corresponding soluble systems. The possible application of immobilized phase-transfer catalysts, as an alternative to soluble ones, is also critically analysed.

The first examples of the application of phase-transfer catalysis (PTC) were described by Jarrousse in 1951 (1), but it was not until 1965 that Makosza developed many fundamental aspects of this technology (2,3). Starks characterized the mechanism and coined a name for it (4,5), whilst Brändström studied the use of stoichiometric amounts of quaternary ammonium salts in aprotic solvents, "ion-pair extraction" (6). In the meantime Pedersen and Lehn discovered crown-ethers (7-9) and cryptands (10,11), respectively.

A lot of convergent knowledge was rapidly acquired in these apparently different fields and an important consequence was that a huge number of structurally different phase-transfer catalysts were made available within a few years; a most significant step was the immobilization of phase-transfer catalysts on a polymer matrix (12-16).

The mechanism of PTC can be represented as follows:

$$R-X + Q^+Y^- \rightleftharpoons R-Y + Q^+X^- \quad \text{(organic phase)}$$
$$\updownarrow \qquad\qquad\qquad \updownarrow$$
$$Y^- + 2M^+ + X^- \quad \text{(aqueous phase)}$$

According to this picture (4,5,17), the reaction occurs in the organic phase in which the anion forms an ion-pair with the lipophilic quaternary cation. In the presence of an excess of nucleophile the reaction follows pseudo-first order kinetics, and the observed rate constant ($\underline{k}_{obsd}$) is linearly related to the molar equivalents of catalysts in the organic phase (Equations 1 and 2).

$$\text{rate} = \underline{k}_{obsd}[RX] \qquad (1)$$

$$\underline{k}_{obsd} = \underline{k}[Q^+Y^-]_{org} \qquad (2)$$

Quaternary onium salts were the first phase-transfer catalysts used; subsequently, a number of compounds (linear polyethers, polypodands, crown-ethers, cryptands, cage-compounds, etc.) were found effective for the anion activation in two-phase systems. These structurally different systems must satisfy at least two fundamental conditions in order to behave as phase-transfer catalysts: i) solubility in the organic phase; ii) steric hindrance around the cationic center leading to a good cation-anion separation within the ion-pair.

A detailed description of the structural requirements and parameters ruling the activity of the most common soluble phase-transfer catalysts was reported recently (18). This account concerns our latest results on phase-transfer catalysis.

Hydration and Reactivity of Anions

Under aqueous-organic two-phase conditions anions dissolved in the organic phase are always accompanied by a certain number of solvating water molecules, hence quaternary salts exist as $Q^+Y^-.nH_2O$ (4,18). The extent of hydration depends essentially on the nature of the anion. For most anions examined (halides and pseudo-halides) the hydration number $\underline{n}$ is in the range of 1-5.

In the case of a few relevant anions we found a good linear correlation between the hydration enthalpies in the gas phase and the hydration numbers when these anions are associated with quaternary cations under PTC conditions (19). The hydration numbers measured for OH^- (11.0±1) and F^- (8.5±1) are in good agreement with those previously extrapolated, i.e. OH^- (10.0) and F^- (9.4) (20,21).

Under PTC liquid-liquid conditions the order of nucleophilic reactivity of halides and pseudo-halides associated with $C_{16}H_{33}P^+Bu_3$ is: $N_3^- > CN^- > Br^- \sim I^- > Cl^- > SCN^-$ (22). This order is anomalous in comparison with the well known reactivity scales both in protic and dipolar aprotic solvents and this behavior is largely due to the specific hydration of the anions in the organic phase. Both the absolute and relative rates are largely modified by working in non-polar organic solvents under anhydrous homogeneous conditions. The absolute rates of reactions increase up to one order of magnitude as the electronegativity of anions increases and their polarizability decreases. The reactivity scale becomes $CN^- > N_3^- > Cl^- > Br^- > I^- > SCN^-$, identical with that found in dipolar aprotic solvents.

Similar behavior was found with cryptates (23), but in this case the absolute rates are higher than those of quaternary salts, in agreement with the better separation between cation and anion.

The behavior of crown-ethers is substantially different (24), as both reaction rates and the reactivity scales ($N_3^- > I^- \sim Br^- > CN^- > Cl^- > SCN^-$), under PTC conditions and in low polarity anhydrous solvents, are very similar. In fact, the removal of the hydration sphere of the anions when passing from two-phase to anhydrous conditions is balanced by a stronger cation-anion interaction.

The influence of hydration on the reactivity of anions is much more evident in the case of OH^-. In the chlorobenzene-aqueous NaOH system the hydration sphere of tetrahexylammonium hydroxide dissolved in the organic phase progressively decreases from 11 to 3.3 water molecules when the base concentration is raised from 15 to 63%. This leads to an enhanced reactivity of OH^- which was measured in the Hofmann elimination (Equation 3). In the examined ranges of NaOH concentrations the reactivity increased up to more than four orders of magnitude (Table I). Although the dehydration of OH^- is not complete, the observed reactivity enhancements are much higher (10^3 times) than those found for halides and pseudo-halides (20,21).

The reactivity differences found in the condensed phase are comparable with those measured by other authors in reactions promoted by $OH^-.(H_2O)_n$ (n = 0-3) in the gas phase (25). All these results confirm that, in anion promoted reactions, the reactivities in the gas phase can be approached by reducing the anion solvation shell in non polar organic solvents.

These results also account for the dramatic effect that an increase of base concentration produces on the rate of the reactions

$$(C_6H_{13})_4\overset{+}{N}\,\overset{-}{OH} \xrightarrow[\text{aq. NaOH, 25°C}]{\text{PhCl}} (C_6H_{13})_3N + \text{hex-1-ene} + H_2O \qquad (3)$$

Table I. Effect of the Specific Hydration on the Basicity of $(C_6H_{13})_4\overset{+}{N}\,\overset{-}{OH}$ in the Hofmann Elimination Reaction

NaOH (%)	Hydration State $\underline{n}$ of $(C_6H_{13})_4\overset{+}{N}\,\overset{-}{OH}\cdot\underline{n}H_2O$	$10^6\underline{k}$ (s^{-1})
15	11.0	0.019 (1)
20	9.0	0.14 (7)
30	5.0	0.74 (39)
40	4.0	26 (1400)
50	3.5	210 (11000)
63	3.3	474 (25000)

promoted by alkali hydroxides under liquid-liquid PTC conditions: e.g. generation and alkylation of carbanions, alkenes isomerization, H-D exchanges in carbon acids, and acid-base equilibria (26).

Finally, when chlorobenzene solutions of tetrahexylammonium salts $(C_6H_{13})_4\overset{+}{N}\,\overset{-}{Y}$ (Y = Cl, MeSO$_3$, Br) were equilibrated with aqueous NaOH solutions, by increasing the NaOH concentration from 15 to 50% the amount of OH$^-$ extracted in the organic phase as Q^+OH^- was found to diminish up to 5 times.

$$(Q^+Y^-)_{org} + (OH^-)_{aq} \rightleftharpoons (Q^+OH^-)_{org} + (Y^-)_{aq} \qquad (4)$$

$$K_{OH/Y}^{sel} = \frac{[Q^+OH^-]_{org}[Y^-]_{aq}}{[Q^+Y^-]_{org}[OH^-]_{aq}} \qquad (5)$$

This behavior is apparently in contrast with the mass effect on the equilibrium (4). It can be explained in terms of selectivity coefficients, $K_{OH/Y}^{sel}$ (Equation 5), which, in the same base concentration range (15-50% NaOH), decrease 41, 100 and 110 times for Y = Cl, MeSO$_3$ and Br, respectively. Therefore, an increase in NaOH concentration not only produces a dramatic enhancement in the reactivity of OH$^-$, but also leads to a decrease in its extractability from the aqueous phase, both effects being due to the dehydration of the hydroxide ion (21).

Lipophilic Macropolycyclic Ligands as Anion Activators

The complexation ability of crown-ethers has been improved by the introduction of secondary donor sites covalently bonded to the macrocyclic ring through a flexible arm, e.g. "lariat ethers" (27). It is also known that in particular conditions crown-ethers can make 2:1 sandwich complexes with the cation (8).

We synthesized bis crown-ethers 1 and 2 in which cations can be well accomodated between the two ligand units, thus giving rise to sandwich complexes (28) which can be expected to have complexation constants higher than those of crown-ethers.

These systems are made lipophilic by the presence of an alkyl chain, and in aqueous-organic two phase conditions are entirely in

$C_{16}H_{33}$

1   n = 2
2   n = 3

the organic phase, but the extent of complexation is modest, even in the presence of highly polarizable anions. Things change dramatically under solid-liquid conditions. A systematic investigation on the complexing ability of ligands 1 and 2 with alkali iodides under solid-liquid conditions showed that the best results are obtained with Na^+ and K^+, the extent of complexation being comparable with that of dicyclohexano-18-crown-6 3. With harder anions (Br^- and Cl^-) 1 and 2 are noticeably better complexing ligands than 3. Accordingly, catalytic activities of 1 and 2 in nucleophilic substitutions on n-octyl methanesulfonate are similar to those found with 3 when I^- is the nucleophile; but 1 and 2 are more efficient catalysts than 3 when Br^- and Cl^- are involved (Table II).

The alkyl substituted proton cryptate $[H^+C(1.1.1, C_{14})]\ Y^-$ 4, can be obtained in satisfactory yields via an hydrogen-bonding templated synthesis (29,30). As for the unsubstituted $[H^+C(1.1.1)]\ Y^-$ cryptate 5, the proton cannot be removed from the intramolecular cavity, so that the species $[H^+C(1.1.1,C_{14})]$ represents a "unique example" of protonated trialkylamine which behaves, in all essential aspects, like a lipophilic tetraalkyl ammonium cation (31). In nonpolar media the high nucleophilic reactivity of the anions associated with cryptate 4 is comparable with that found for the best anion activators like $[K \overset{+}{\subset} (2.2.2,C_{14})]$ Y^- 6 and $(C_8H_{17})_4\overset{+}{N}Y^-$ 7 (Table III).

$$4 \quad R = C_{14}H_{29}-n$$
$$5 \quad R = H$$

Table II. Catalytic Activity of Sandwich Ligands 1 and 2 in Nucleophilic Substitutions on n-Octyl Methanesulfonate with Alkali Halides (MY) under Solid-Liquid Conditions in Toluene at 50°C

MY	$\underline{k}_{obsd} \times 10^6, s^{-1}$		
	1	2	3
NaCl	3.4	3.1	negligible
KCl	2.8	2.8	negligible
NaBr	30	28	5.0
KBr	20	21	5.3
NaI	240	192	255
KI	82	170	160

Table III. Second-Order Rate Constants for Nucleophilic Substitutions on n-Octyl Methanesulfonate by Anions (Y^-) Associated with $[H\overset{+}{\subset}(1.1.1,C_{14})]$ 4, $[K\overset{+}{\subset}(2.2.2,C_{14})]$ 6, $(C_8H_{17})_4\overset{+}{N}$ 7 and $[Na\overset{+}{\subset}(Cage Ligand)]$ 8 in Anhydrous Chlorobenzene at 60°C

Y^-	$10\dfrac{k}{2}, M^{-1}s^{-1}$			
	4	6	7	8
Cl$^-$	1.9	5.1	3.7	–
Br$^-$	1.1	3.7	2.0	2.8
I$^-$	0.60	0.87	0.68	0.86
N$_3^-$	7.0	15.0	15.6	–

Complexes 8 and 9 of cage ligands derived from 1,7-dioxa-4,10-diazacyclododecane can be used as anion activators both in stoichiometric and catalytic reactions and their activity is comparable with that found for lipophilic $[K^+ \subset (2.2.2,C_{14})]Y^-$ cryptates 6 (28,32) (Table III).

Both quaternary onium salts and cation complexes of lipophilic multidentate ligands (crown-ethers and cryptands) have been used as catalysts in two-phase systems in the presence of base (OH$^-$, F$^-$, etc.). However, under these conditions, the lack of chemical stability of quaternary salts and the very low complexation constants of multidentate ligands (especially crown-ethers) make all these systems barely effective in the activation of such anions.

A peculiar feature of sodium complexes 8 and 9 is their exceptionally high stability, which allows a high catalytic efficiency

$$R-\left[\text{cage-Na}^+\right] Y^- \quad \begin{array}{l} 8 \quad R = PhCH_2 \\ 9 \quad R = C_{16}H_{33}\text{-}\underline{n} \end{array}$$

also in reactions promoted by very hydrophilic anions. For example, using catalytic amounts of 8 or 9 (Y = OCH$_3$) $\underline{n}$-octyl methanesulfonate was converted to the corresponding fluoride with aqueous KF (75% yield, 1h at 110°C) and, in an oxygen atmosphere, diphenylmethane was converted to benzophenone with 15M aqueous NaOH (95% yield, 1h at 70°C) (32).

Phase-Transfer Catalysts Immobilized into a Polymeric Matrix

Catalysts have been bonded to insoluble polymers to allow, in principle, an appreciable simplification of PTC: the catalyst represents a third insoluble phase which can be easily recovered at the end of the reaction by filtration, thus avoiding tedious processes of distillation, chromatographic separation and so on. This is of potential interest mainly from the industrial point of view, due to the possibility of carrying on both discontinuous processes with a dispersed catalyst and continuous processes with the catalyst on a fixed bed. This technique was named "triphase catalysis" by Regen (13,33,34).

The direct quaternization of chloromethylated polystyrenes by tertiary amines or phosphines represents the easiest way to obtain polymer-supported quaternary onium salt (12,13). A lipophilic character of quaternary cation and a topology allowing sufficient cation-anion separation also play an important role (35,36). A linear spacer chain (of about 10 carbon atoms) between the catalytic site and the polymer backbone substantially increases the reaction rates. The loading of quaternary onium groups also affects catalytic efficiency, the influence being different for directly bonded and spaced groups, e.g. 10 and 11, respectively (37).

The catalytic activity of 10 decreases by about one order of magnitude going from 10 to 60% ring substitution and log k_{obsd} is linearly related to the loading. With spaced catalysts 11 the observed rate constants are 1.7-3.1 times higher than those of 10 in the range of 10-30% ring substitution, whereas they are 4.1-10 times higher at 60% ring substitution. This behavior can be explained by

P—〔Ph〕—$CH_2\overset{+}{P}Bu_3$ Cl^- P—〔Ph〕—$CH_2NHCO(CH_2)_{10}\overset{+}{P}Bu_3$ Br^-

10a	(6.1)		11a	(5.8)
b	(10.6)		b	(8.7)
c	(28.8)		c	(25.4)
d	(60.0)		d	(57.3)

% ring substitution in parenthesis

the combination of two opposite effects: i) a polarity increase at the catalytic site by increasing the percentage ring substitution, thus lowering the anionic reactivity; ii) a polarity decrease by inserting a spacer chain, the reason being the intrinsic lipophilicity of the latter and a more even distribution of catalytic sites within the polymer matrix (37).

A huge number of experimental results indicate that phase-transfer catalyzed reactions follow an identical mechanism both in the presence of soluble and polymer-supported catalysts (35,38). Reactions should occur in the organic shell surrounding the catalytic site. Anions exchange at water-organic interface, with the inorganic cation M^+ and the polymer-supported cation (P)~Q^+ as counterions in the aqueous and organic phase, respectively. When structural requirements are ensured and diffusive factors are minimized, the efficiency of polymer-supported quaternary salts is only slightly lower than that found with similar soluble catalysts.

Suitably functionalised crown-ethers and cryptands have been synthesized and reacted with chloromethyl polystyrene. Initially macrocyclic units were immobilized on polymers through an amino linkage (14,15,39). Recently, a more extensive study was accomplished with crown-ethers immobilized through an ether bond, 12,13, (40).

Three parameters mainly influence the catalytic activity of 12 and 13: i) electronegativity and polarizability of the nucleophile; ii) the percent ring substitution; iii) the presence of a spacer chain. With soft nucleophiles (e.g. I$^-$) log k_{obsd} linearly decreases about 5 times on going from 5 to 60% ring substitution, whereas with

$$\text{P} - \text{C}_6\text{H}_4 - \text{CH}_2\text{O(CH}_2)_n$$

n = 0	12a	(6.8)	n = 9	13a	(4.5)
	b	(10.6)		b	(7.6)
	c	(28.7)		c	(30.6)
	d	(62.3)		d	(62.0)

% ring substitution in parenthesis

harder anions (e.g. Br$^-$) log k_{obsd} shows a maximum at 30% ring substitution. In all cases spaced catalysts 13 were 2-4 times more reactive than directly bonded ones 12 (40).

In order to compare the catalytic activity of polymer-supported crown-ethers with the corresponding phosphonium salts we defined efficiency coefficients, E_{pol} (CE/QY), (Equation 6).

$$E_{pol}(CE/QY) = k_{obsd} \ 12/k_{obsd} \ 10 \text{ or } k_{obsd} \ 13/k_{obsd} \ 11 \quad (6)$$

Efficiency coefficients of catalysts 12 and 13 mainly depend on the nature of the nucleophile, and are in the range 2.1-3.3 and 0.07-0.74 for I$^-$ and Br$^-$, respectively (Table IV). This trend is in good agreement with that found for soluble catalysts (40).

With soft anions crown-ethers are more efficient than quaternary salts, the reverse being observed when less polarizable nucleophiles are used. This is explained by the different extent of complexation of crown-ethers which depends not only on the complexed cation, but also on the anionic counterpart. Swelling and hydration measurements of polymer-supported crown-ethers in toluene/aqueous KY showed that the content of water in the imbibed solvent increases with the loading. This leads to a progressive polarity increase within the polymers and to a better crown-ether complexing capability, more relevant for hard anions (40).

Table IV. Catalytic Efficiency, E_{pol} (CE/QY), of Polymer-Supported Crown-Ethers 12a-d and 13b-d in Nucleophilic Substitutions on n-Octyl Methanesulfonate in Toluene-H_2O at 60°C

Catalyst	% Ring Substitution	E_{pol} (CX/QY)	
		Y = I	Y = Br
12a	6.8	2.1	0.07
12b	10.6	2.1	0.10
12c	28.7	2.3	0.22
12d	62.3	3.3	0.36
13b	7.6	2.5	0.15
13c	30.6	3.1	0.74
13d	62.0	2.5	0.72

A systematic analysis of factors ruling the catalytic activity of polymer-supported cryptands turned out to be impossible due to the difficulties of synthesizing a homogeneous series of catalysts in a wide enough range of loading. Immobilized cryptands can be easily prepared by the condensation of functionalized cryptands with lightly loaded chloromethyl polystyrenes ($\leq$ 1 mequiv Cl/g) (39,41). However, with polystyrenes having higher contents of chloromethyl groups, attack of cryptands was accompanied by extensive structural

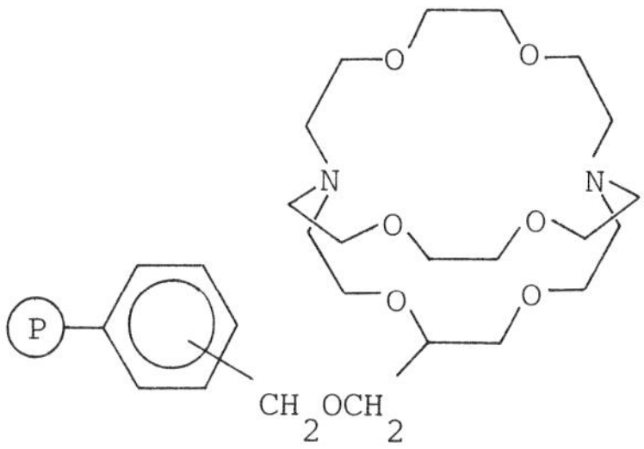

14a (2.8)
b (6.0)

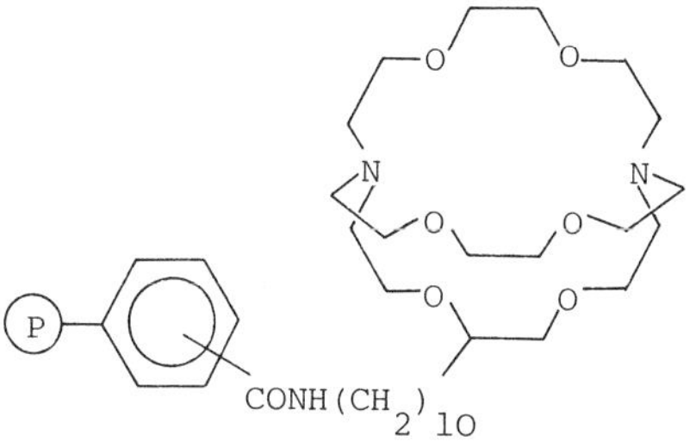

15a (5.1)
b (20.7)

% ring substitution in parenthesis

modifications of the bicyclic ligand (28). It is more likely that the already linked cryptands undergo quaternization at bridgehead nitrogens by neighbouring chloromethyl groups, followed by Hofmann degradation and opening of the bicyclic structure (28).

Catalysts with higher cryptand unit contents (up to 20% ring substitution) could be obtained by the condensation of carboxylated polystyrenes with aminoalkyl substituted cryptands, catalysts 15, (28).

As previously emphasized, the catalytic efficiency of cryptands is generally higher than that of the corresponding quaternary salts and crown-ethers, both for soluble and polymer-supported systems. A few representative values of efficiency coefficients E_{pol} (CRY/QY) (Equation 7), are reported in Table V. They are in agreement with the peculiar property of cryptands whose complexation constants are always very high and, differently from crown-ethers, are almost independent of the nature of the anion.

$$E_{pol}(CRY/QY) = k_{obsd}\ 14/k_{obsd}\ 10 \text{ or } k_{obsd}\ 15/k_{obsd}\ 11 \qquad (7)$$

Table V. Catalytic Efficiency, E_{pol} (CRY/QY), of Polymer-Supported Cryptands 14 and 15 in Nucleophilic Substitutions on n-Octyl Methanesulfonate in Toluene-H_2O at 60°C

		E_{pol} (CRY/QY)	
Catalyst	% Ring Substitution	Y = I	Y = Br
14a	2.8	4.0	2.7
14b	6.0	2.7	1.6
15a	5.1	1.3	0.7
15b	20.7	1.4	0.6

Conclusions

A number of organic molecules capable of efficiently operating as phase-transfer catalysts is now available. The reaction mechanism both for soluble and polymer-supported systems is completely understood and the factors ruling the reactivity are recognised. The drawback of soluble catalysts is their difficult separation from the reaction products which in the case of the expensive macropolycyclic ligands imposes severe limitations in their use on a large scale. The cheap and easy to synthesize ammonium quaternary salts, providing they are stable under the reaction conditions, represent the catalysts of choice.

Immobilization of phase-transfer catalysts on polymeric matrices avoids the problem of separating and recycling the catalysts. In this case the chemical stability of the immobilized catalyst becomes very important: quaternary salts often decompose under drastic reaction conditions whereas polydentate ligands are always stable. However, the difficult synthesis of cryptands, despite their high catalytic efficiency, can hardly justify their use. Synthesis of crown-ethers is much easier, but catalytic efficiencies are often too low.

Several industrial processes use phase-transfer techniques with soluble catalysts, mostly for fine chemical productions (42,43). It is easy to believe that this technology will find a greater application in the near future, perhaps with the use of polymer-supported catalysts.

Literature Cited

1. Jarrousse, J. Compt. Rend. Acad. Sci. Paris 1951, 232, 1424.
2. Makosza, M.; Serafin, B. Rocz. Chem. 1965, 39, 1223.
3. Makosza, M. Pure Appl. Chem. 1975, 43, 439.
4. Starks, C.M. J. Am. Chem. Soc. 1971, 93, 195.
5. Starks, C.M.; Owens, R.M. J. Am. Chem. Soc. 1973, 95, 3613.
6. Brändström, A. "Preparative Ion Pair Extraction"; Lakemedel: Apotekarsocieteten, AB Hassle, 1974.
7. Pedersen, C.J. J. Am. Chem. Soc. 1967, 89, 2495.
8. Pedersen, C.J. J. Am. Chem. Soc. 1967, 89, 7017.
9. Pedersen, C.J., Frensdorff, H.K. Angew. Chem., Int. Ed. Engl. 1972, 11, 16.
10. Dietrich, B.; Lehn, J.-M.; Sauvage, J.P. Tetrahedron Lett. 1969, 2885.
11. Lehn, J.-M Structure and Bonding 1973, 16, 1.
12. Regen, S.L. J. Am. Chem. Soc. 1975, 97, 5956.
13. Regen, S.L. Angew. Chem., Int. Ed. Engl. 1979, 18, 421.
14. Cinquini, M.; Colonna, S.; Molinari, H.; Montanari, F.; Tundo, P. J. Chem. Soc., Chem. Commun. 1976, 394.
15. Molinari, H.; Montanari, F.; Tundo, P. J. Chem. Soc., Chem. Commun. 1977, 639.
16. Brown, J.M.; Jenkins, J.A. J.Chem.Soc.,Chem. Commun. 1976, 458.
17. Landini, D.; Maia, A.; Montanari, F. J. Chem. Soc., Chem. Commun. 1977, 112.
18. Montanari, F.; Landini, D.; Rolla, F. Top. Curr. Chem. 1982, 101, 147, and references therein.
19. Landini, D.; Maia, A.; Montanari, F. J. Am. Chem. Soc. 1984, 106, 2917.
20. Landini, D.; Maia, A. J. Chem. Soc., Chem. Commun. 1984, 1041.
21. Landini, D.; Maia, A.; Rampoldi, A., unpublished results.

22. Landini, D.; Maia, A.; Montanari, F. J. Am. Chem. Soc. 1978, 100, 2796.
23. Landini, D.; Maia, A.; Montanari, F.; Tundo, P. J. Am. Chem. Soc. 1979, 101, 2526.
24. Landini, D.; Maia, A.; Montanari, F.; Pirisi, F.M. J. Chem. Soc., Perkin Trans. 2 1980, 46.
25. Bohme, D.K.; Mackay, G.I. J. Am. Chem. Soc. 1981, 103, 978.
26. Dehmlow, E.V.; Dehmlow, S.S. "Phase Transfer Catalysis"; Verlag Chemie: Weinheim, 1983, 2nd Ed.
27. Dishong, D.M.; Diamond, C.J.; Cinoman, M.I.; Gokel, G.W. J. Am. Chem. Soc. 1983, 105, 586.
28. Anelli, P.L.; Montanari, F.; Quici, S., unpublished results.
29. Annunziata, R.; Montanari, F.; Quici, S.; Vitali, M.T. J. Chem. Soc., Chem. Commun. 1981, 777.
30. Anelli, P.L.; Montanari, F.; Quici, S. J. Org. Chem. 1985, 50, 3453.
31. Landini, D.; Maia, A.; Montanari, F.; Quici S. J. Org. Chem. 1985, 50, 117.
32. Anelli, P.L.; Montanari, F.; Quici, S. J. Chem. Soc., Chem. Commun. 1985, 132.
33. Ford, W.T.; Tomoi, M. Adv. Polym. Sci. 1984, 55, 49.
34. Akelah, A.; Sherrington, D.C. Chem. Rev. 1981, 81, 557.
35. Molinari, H.; Montanari, F.; Quici, S.; Tundo, P. J. Am. Chem. Soc. 1979, 101, 3920.
36. Tomoi, M.; Ford, W.T. J. Am. Chem. Soc. 1981, 103, 3821.
37. Anelli, P.L.; Montanari, F.; Quici S. J. Chem. Soc., Perkin Trans. 2 1983, 1827.
38. Montanari, F.; Quici, S.; Tundo, P. J. Org. Chem. 1983, 48, 199.
39. Montanari, F.; Tundo, P. J. Org. Chem. 1981, 46, 2125.
40. Anelli, P.L.; Czech, B.; Montanari, F.; Quici, S. J. Am. Chem. Soc. 1984, 106, 861.
41. Anelli, P.L.; Quici, S. Synthesis, 1985, 1070.
42. Reuben, B.; Sjöberg, K. Chemtech 1981, 315.
43. Ford, W.T. Chemtech 1984, 436.

RECEIVED July 26, 1986

Chapter 7

Efficient Asymmetric Alkylations via Chiral Phase-Transfer Catalysis: Applications and Mechanism

U.-H. Dolling, D. L. Hughes, A. Bhattacharya, K. M. Ryan, S. Karady, L. M. Weinstock, and E. J. J. Grabowski

Merck Sharp & Dohme Research Laboratories, Merck & Co., Inc., Rahway, NJ 07065

> Catalytic asymmetric methylation of 6,7-dichloro-5-methoxy-2-phenyl-1-indanone with methyl chloride in 50% sodium hydroxide/toluene using N-(p-trifluoromethylbenzyl)cinchoninium bromide as chiral phase transfer catalyst produces (S)-(+)-6,7-dichloro-5-methoxy-2-methyl-2-phenyl-1-indanone in 94% ee and 95% yield. Under similar conditions, via an asymmetric modification of the Robinson annulation employing 1,3-dichloro-2-butene (Wichterle reagent) as a methyl vinyl ketone surrogate, 6,7-dichloro-5-methoxy-2-propyl-1-indanone is alkylated to (S)-(+)-6,7-dichloro-2-(3-chloro-2-butenyl)-2,3-dihydroxy-5-methoxy-2-propyl-1-inden-1-one in 92% ee and 99% yield. Kinetic and mechanistic studies provide evidence for an intermediate dimeric catalyst species and subsequent formation of a tight ion pair between catalyst and substrate.

Efficient asymmetric alkylations have been a long-sought goal in organic synthesis. In the recent literature three different approaches have been used for asymmetric alkylations of ketones: 1) chiral alkylating reagents (2, 3); 2) chiral auxiliaries (4-9); and 3) chiral catalysts (10-12). Two examples of the first approach are shown in Figure 1. Murphy observed an asymmetric induction of 8% during the alkylation of 2,4,6-trimethylphenol with allyl (+)-camphor-10-sulphonate. Duhamel methylated the Schiff base of methyl glycinate with 1,2,4,6-bis-(0-isopropylidene)-3-methoxy-sulfonyl-α-D-glucose (A*) and produced S-alanine in 65% yield, 40% enantiomeric excess (ee).

The use of chiral auxiliaries has been developed into elegant three-step sequences to achieve high ee's (Figure 2). In the general scheme a ketone is derivatized with a chiral amine. Low temperature lithiation and alkylation followed by hydrolysis produces the alkylated ketone in moderate to excellent ee's. The auxiliaries most often used are (S)-valine tert-butyl ester (Koga), 1-amino-2-methoxymethylpyrrolidine (Enders) and (S)-2-amino-1-

0097-6156/87/0326-0067$06.00/0
© 1987 American Chemical Society

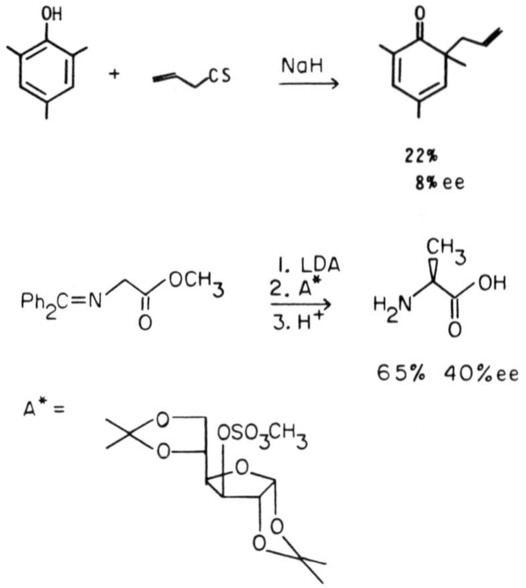

Figure 1. Asymmetric alkylations with chiral alkylating reagents.

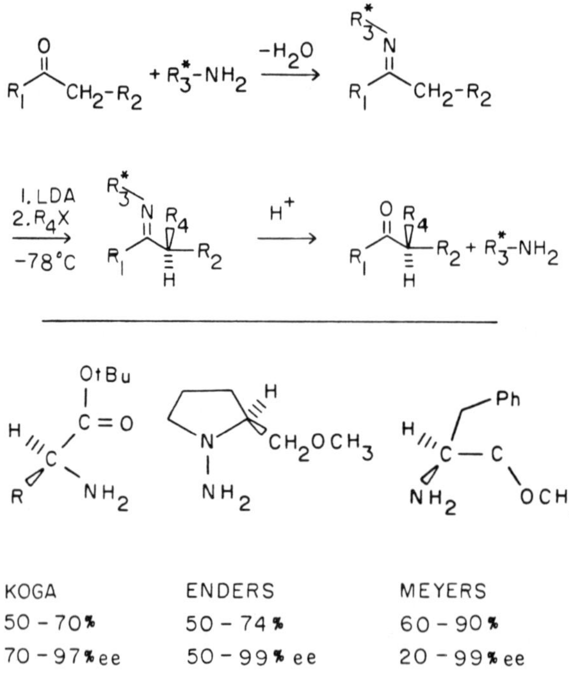

Figure 2. Asymmetric alkylations with chiral auxiliaries.

KOGA
50 - 70%
70 - 97% ee

ENDERS
50 - 74%
50 - 99% ee

MEYERS
60 - 90%
20 - 99% ee

methoxy-3-phenylpropane (Meyers). The disadvantage of both methods with respect to an economical process is that the chiral auxiliaries are used in stoichiometric amounts and have to be recycled. Chiral phase-transfer-mediated alkylations offer a potentially simple, one-step solution to this problem. The first reports of such alkylations using cyclic β-keto esters as substrates and ephedrinium or N-benzylquininium halides as catalyst claimed ee's of only 15% (10-12) (Figure 3). Higher catalytic asymmetric inductions were achieved in epoxidations (13,14) and Michael addition reactions using chiral Crown ethers (15) and cinchona alkaloids or their quaternized derivatives (16-19) (Figure 4).

Our interest in chiral alkylations arose relative to the diuretic drug candidate Indacrinone 7 which had reached the stage of clinical evaluation (20-23). It was originally prepared as a racemate (Scheme 1), but resolution and further testing revealed that the S-(+)-enantiomer showed uricosuric activity, while the diurectic activity resided mainly in the R-(-)-enantiomer. Both activities are desirable, but the R-enantiomer is much more active. To balance the therapeutic effects a 90/10 ratio of the S/R-enantiomers was proposed as the drug candidate. Since an achiral phase-transfer-catalyzed (PTC) methylation was a key step in the synthesis of the racemate our attention naturally turned to incorporating chirality into this step, despite the dismal outlook for success suggested by the literature. Herein we will discuss the preliminary development of the catalytic chiral phase transfer methylation and some unusual mechanistic aspects of this reaction. A concluding section will deal with the application of this methodology to a new asymmetric Robinson annulation.

Chiral PTC Methylation

Preliminary work with N-alkyl derivatives of the various cinchona alkaloids (18, 19) readily established N-benzylcinchoninium chloride as potentially one of the most effective catalysts, although ee's were in the modest range (20-30%). Subsequent development of the reaction with respect to each of the reaction variables proved critical to the success of the chiral phase transfer approach (Figure 5). Nonpolar solvents such as toluene or benzene produced higher ee's than polar solvents such as methylene chloride or methyl tert-butyl ether. The ee's increased with higher dilution, increasing NaOH concentrations, faster agitation and lower temperature. The catalyst concentration (10-50 mole % range) controlled the rate of reaction but had little effect on the ee. As catalyst counterion, chloride and bromide produced similar results, whereas iodide decreased the ee substantially. As alkylating agent CH_3Cl gave by far the best selectivity relative to CH_3Br, CH_3I or dimethyl sulfate.

CPK molecular models, the single crystal x-ray structure, and molecular modeling studies of the N-benzylcinchoninium ion suggest that a preferred conformation may be that in which the quinoline ring, the C_9-O bond, and the N-benzyl group all lie in one plane (Figure 6). The anion of 5 also has an almost planar structure with the negative charge delocalized into the 2-phenyl ring. Both

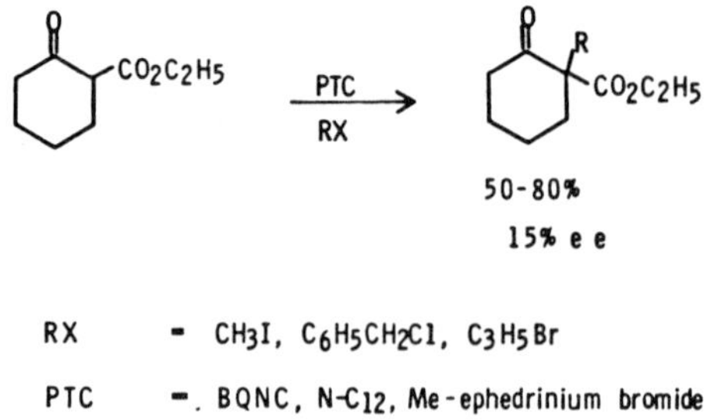
CO2C2H5
PTC
RX
R
CO2C2H5
50-80%
15% e e
RX - CH3I, C6H5CH2Cl, C3H5Br
PTC -. BQNC, N-C12, Me-ephedrinium bromide

Figure 3. Asymmetric alkylations with chiral phase transfer catalysts.

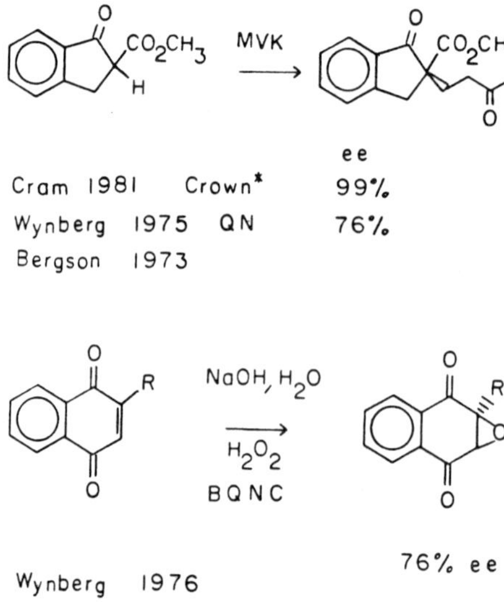
CO2CH3
H
MVK
CO2CH3
ee
Cram 1981 Crown* 99%
Wynberg 1975 QN 76%
Bergson 1973
R
NaOH, H2O
H2O2
BQNC
R
76% ee
Wynberg 1976

Figure 4. Catalytic asymmetric Michael addition and epoxidation.

7. DOLLING ET AL. Efficient Asymmetric Alkylations   71

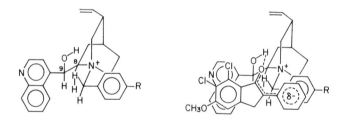

X: I, Br, Cl, OSO₃CH₃

PTC*:

R_2 = OCH₃, H   R = OCH₃, CH₃, H, F, Cl, CF₃

Y = I⁻, Br⁻, Cl⁻

Figure 5. Variations of the chiral phase transfer methylation of indanone 5.

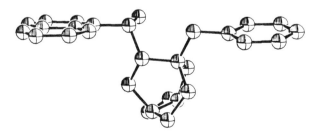

Figure 6. (a) Ion pairing between indanone anion and benzylcinchoninium cation.

Figure 6. (b) ORTEP structure of the catalyst cation rotated 90° to show the planar catalytic surface.

molecules in their nearly planar conformations fit naturally on top of each other providing π-interaction between the benzyl group of the catalyst and the 2-phenyl of anionic 5 on the one side and between the quinoline and methoxydichlorobenzene moieties on the other. The C_9-hydroxyl provides a directional handle for the ionic attraction via hydrogen bonding to the indanone anion. The CH_3Cl can only alkylate from the front side and form the (S)-(+)-6,7-dichloro-5-methoxy-2-methyl-2-phenyl-1-indanone.

The ion pairing between the enolate of 5 and the catalyst should make the asymmetric induction sensitive to the electronic effects of substituents on the N-benzyl group. A Hammett plot of log ee/ee$_0$ vs the substituent constant σ of the para N-benzyl substituted catalysts (R = CH_3O, CH_3 H, F, Cl, CF_3) gave a reaction constant of ρ = 0.21 ± 0.02 with ee's in the range of 60% to 92% demonstrating that substituents with increasing electron-withdrawing power improve the catalyst selectivity (Figure 7).

These initial studies clearly demonstrated that N-(p-trifluoromethylbenzyl)cinchoninium bromide (p-CF_3BCNB) was the catalyst of choice in combination with CH_3Cl as methylating agent. In a typical reaction 75 ml of toluene, 15 ml of 50% aqueous NaOH, 6 mmole of indanone 5, 0.6 mmole of catalyst, and 42 mmole of CH_3Cl were charged under nitrogen to the reactor (300 ml stirred autoclave, Autoclave Engineers). The reaction was then stirred at 20°C and 1200 rpm for 18 hours resulting in a 95% yield, 92% ee, of 6. Further conversion by established procedures (Scheme 1) produced 8 in 72% isolated yield and >99% ee. An example of the ee analysis of 6 by NMR is shown in Figure 8 ($CDCl_3$, tris[3-(heptafluoropropylhydroxymethylene)-d-camphorato]europium (III)). The relative intensities of the 2-methyl singlets were measured and compared with authentic standards (1).

Mechanistic Aspects

Dehmlow (24) published a cautionary note questioning the early literature results of chiral phase transfer alkylations. He demonstrated that under basic conditions chiral β-hydroxy ammonium catalysts undergo Hofmann elimination to give optically active oxiranes. These may contaminate the products and generate apparent optical activity. Indeed we find small amounts of the corresponding oxirane 12 in our reaction mixtures (Figure 9). However, under basic alkylating conditions the major decomposition product arises via alkylation of the alkoxide 9. The resulting methyl ether 10 readily undergoes E2 elimination forming the enol ether 11. Oxirane 12, methyl ether 10, and enol ether 11 were independently prepared and characterized. Under reaction conditions the methyl ether 10 is rapidly converted to the enol ether. Both oxirane and enol ether are methylated with CH_3I under reaction conditions to form the new quaternary ammonium salts 13 and 14 which act as phase transfer catalysts and produce racemic 2-methylindanone 6. This can account in part for the low asymmetric induction using CH_3I as alkylating reagent. With CH_3Br, alkylation of tertiary amines 11 and 12 is slow relative to the alkylation of indanone 5. CH_3Cl does not

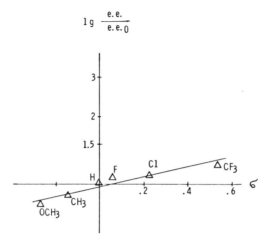

Figure 7. Hammett plot.

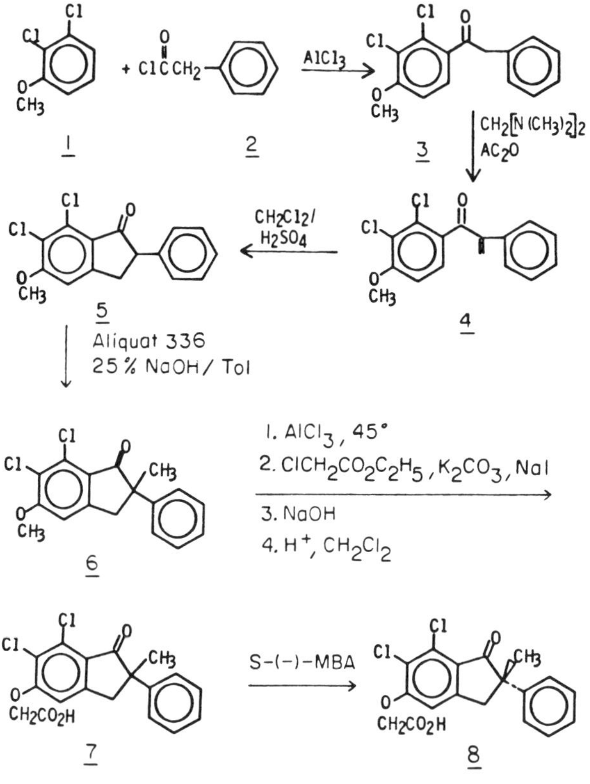

Scheme 1. Racemic synthesis and resolution of indacrinone.

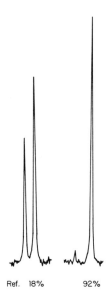

Ref.   18%   92%

Figure 8. NMR assays of a synthetic mixture (18% ee) of (+)- and (-)-enantiomers and a reaction product mixture (92% ee).

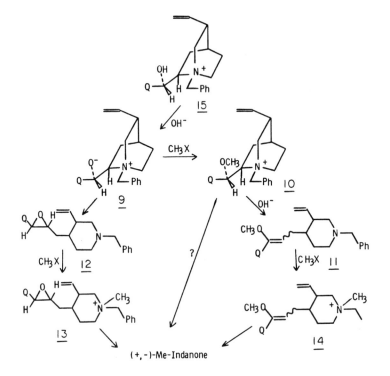

Figure 9. Catalyst decomposition pathways.

alkylate amines under the reaction conditions. Consistent with the low reactivity of CH_3Br and CH_3Cl towards 11 and 12 is the observation that the ee is constant throughout the reaction. Initial experiments with the preferred catalyst and alkylating agent (Table I) demonstrated some unusual behavior of the chiral alkylation: a) at constant catalyst concentration the ee will dramatically decrease from 90% to 0% with increasing indanone concentration (entries 1 to 3); b) the reaction is zero-order in indanone; c) for a twofold increase of indanone 5 the pseudo zero-order rate constant for the (+)-isomer formation decreased roughly fivefold; (d) at constant indanone concentration the ee will increase with increasing mole % catalyst (entries 3, 4, 5 and 7, 2, 6); (e) at a constant molar ratio of indanone to catalyst the ee will increase with higher dilution (entries 5, 6 and 1).

Table I. Effect of Catalyst/Indanone Ratio on
Rate and Selectivity of the Chiral Methylation[1]

	5 mmole	p-CF3BCNB mmole	mole % of 5	ee (+) %	k (+)-enantiomer mole $l^{-1}s^{-1} \times 10^6$
1	2.5	0.25	10	90	2.4
2	5	0.25	5	69	0.4
3	10	0.25	2.5	0	0.07
4	10	0.50	5	70	0.7
5	10	1.0	10	82	5.8
6	5	0.50	10	87	4.0
7	5	0.125	2.5	0	0.03

1. Reaction conditions: 125 ml toluene, 25 ml 50% NaOH, 70 mmole CH_3Cl, 1200 rpm, 20°C, autoclave.

To help understand these findings we examined the chiral alkylation in detail. The reaction profile in Figure 10 displays the concentrations of indanone 5 and catalyst in the toluene layer in mg/ml and the % indanone 5 remaining in the total mixture, the balance being the product 2-methylindanone 6. The graph reveals the following important points: 1) the methylation proceeds slowly during the first 6 to 7 hours; 2) during this "induction period" the concentration of the indanone in toluene initially drops to less than 0.2 mg/ml; and 3) not until the indanone is removed from the toluene do the catalyst concentration and the rate of methylation increase. This unique behavior was found to be caused by two factors: solid enolate and catalyst dimer formation. This is illustrated by the following experiments. When a 16 mM solution of indanone 5 in toluene is stirred in the absence of any catalyst with 50% aqueous NaOH, the indanone is deprotonated at the interface to form the sodium enolate which precipitates, thus reducing the concentration of 5 in toluene (Figure 11). The enolate was identified by IR. The carbonyl absorption at 1700 cm^{-1} was replaced by two new peaks at 1525 cm^{-1} and 1565 cm^{-1} due to the enolate anion (25). The precipitation of the sodium enolate is slower with 40% NaOH. With 30% NaOH the enolate does not precipitate and, therefore, the indanone concentration remains high. The reaction profile in Figure 10 shows that the catalyst concentration is initially very low and

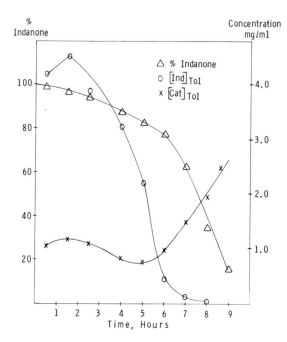

Figure 10. Chiral phase-transfer reaction profile.

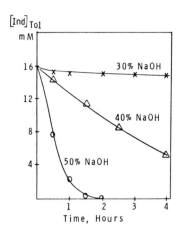

Figure 11. Interfacial deprotonation of indanone **5** in toluene/aqueous NaOH.

does not increase until the enolate formation is complete. Examining the catalyst solubility we surprisingly found the catalyst is not soluble in toluene or in 50% NaOH alone (<10^{-5} M). How, then, does p-CF$_3$BCNB act as a phase transfer catalyst? p-CH$_3$BCNB does dissolve in toluene in the presence of 50% NaOH as a dimer of the catalyst 15 and its zwitterionic oxide 9 (Figure 9) based on the following experimental evidence. This catalyst solution contains one mole of base and one mole of bromide per two moles of catalyst. Titration with acid does not generate an equivalent amount of water. A quench of the catalyst solution with CH$_3$I generates a 50/50 mixture of catalyst 15 and methyl ether 10. Addition of indanone 5 to the catalyst solution results in deprotonation of 5 by the zwitterion. Since the zwitterion is now protonated the catalyst is no longer soluble and its concentration in the toluene drops. In the reaction profiled in Figure 10 only a low level of the catalyst enters the toluene layer as long as indanone is present but when all the indanone has been deprotonated the concentration of the catalyst will increase.

These observations showed that the reaction can be simplified by preformation of the indanone enolate in toluene/50% NaOH and subsequent addition of catalyst and CH$_3$Cl (Figure 12). This eliminates the "induction period" and most importantly the high sensitivity of rate and ee to the catalyst/indanone ratio. Detailed kinetic measurements on this preformed enolate methylation in toluene/50% NaOH determined that the reaction is 0.55 order in catalyst. This is consistent with our finding that the catalyst goes into solution as a dimer which must dissociate prior to complexation with the indanone anion. If the rate has a first order dependence on the monomer, the amount of monomer is very small, and the equilibration between dimer and monomer is fast, then the order in catalyst is expected to be 0.5. The 0.5 order in catalyst is not due to the preformation of solid sodium indanone enolate but is a peculiarity of this type of chiral catalyst. When Aliquat 336 is used as catalyst in this identical system the order in catalyst is 1. Finally, in the absence of a phase transfer catalyst less than 2% methylation was observed in 95 hours.

When the chiral methylation is carried out with 30% aqueous NaOH the indanone is deprotonated at the interface but does not precipitate as the sodium enolate (Figure 11). In this system there are 3 to 4 molecules of H$_2$O per molecule of catalyst available while in the 50% NaOH reactions the toluene is very dry with only 1 molecule of H$_2$O available per catalyst molecule thus forcing the formation of tight ion pairs. Solvation of the ion pairs in the toluene/30% NaOH system should decrease the ee which we indeed observe with an optimum 78% versus 94% in the 50% NaOH reaction. In the 30% NaOH reactions the ee decreases from 78% to 55% as the catalyst concentration increases from 1 mM to 16 mM (80 mM 5, 560 mM CH$_3$Cl, 20°C). Based on these ee's rates of formation of (+)-enantiomer and racemic product can be calculated. When the log of these rates are plotted versus the log of catalyst concentrations (Figure 13) we find an order of about 0.5 in the catalyst for the chiral process similar to that found using 50% NaOH consistent with a dimer-monomer pre-equilibrium. The order in catalyst for the

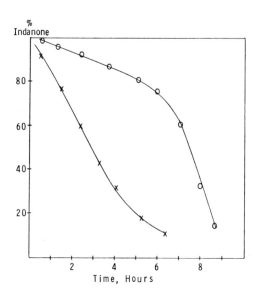

Figure 12. Rate of methylation: initial (O) versus delayed (X) catalyst charge.

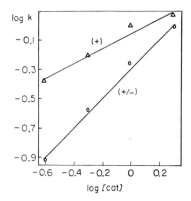

Figure 13. Order in catalyst for racemic (1.0) and chiral (0.5) methylations using 30% NaOH.

racemate formation is 1.0 suggesting a different pathway, possibly via dimer catalysis.

Chiral Robinson Annulation

During the elaboration of the chiral PTC methylation the need arose for a chiral synthesis of the tricyclic compound 20, an intermediate for a drug candidate under development (26). The racemate is readily prepared by base catalyzed methyl vinyl ketone (MVK) addition to 2-propylindanone 17 followed by base or acid catalyzed aldol condensation (Figure 14). We recognized that the synthesis can readily be adapted to the chiral phase transfer alkylation by use of 1,3-dichloro-2-butene (Wichterle reagent) as an MVK surrogate for the Michael addition and overall Robinson annulation (27-29). Three significant differences are present in this application: a) substitution of the indanone with a propyl rather than a phenyl group; b) a larger alkylating reagent of different reactivity; and most importantly, c) the opposite sterochemistry at the C_2 carbon is required. Under the reaction conditions described earlier, alkylation of 17 using p-CF_3BCNB as catalyst produced (S)-(+)-6,7-dichloro-2-(3-chloro-2-butenyl)-2,3-dihydro-5-methoxy-2-propyl-1-inden-1-one 19 in 99% yield and 92% ee. Hydrolysis (0°C, 1.5 h) and cyclization (70°C, 1 h) in concentrated sulfuric acid produced 20 in 96% overall yield. As expected, based on the tight ion pair theory previously proposed, the absolute configuration was opposite to the desired one shown in Figure 14. Initial experiments with N-(p-trifluoromethylbenzyl)cinchonidinium bromide as catalyst resulted only in modest ee's and low yields of the opposite R-(-)-enantiomer. Optimization of the concentrations of indanone, 1,3-dichloro-2-butene and catalyst (N-p-trifluoromethylbenzyl)cinchonidinium chloride) improved the yield to 97%, 76% ee, using 30 mole % of the catalyst.

The decrease in enantioselectivity was not unexpected because cinchonidine is a diastereomer and not an enantiomer of cinchonine with only two (C_8 and C_9) of the four chiral centers inverted. Study of space filling models suggests that the vinyl group in the cinchonine catalyst blocks the hydroxyl from the backside thus making it available for hydrogen bonding only from the front, the asymmetric catalyst surface. The vinyl group blocking of the hydroxyl is absent in the cinchonidine catalyst 16 which may confer greater degrees of freedom to the system and hence lower ee's (Figure 15).

In summary, we have demonstrated the first efficient enantioselective alkylation via phase transfer catalysis. This alkylation was expanded to include an enantioselective Robinson annulation. The methodology was developed for the preparation of either enantiomer. Finally, our kinetic studies have provided additional mechanistic insight into the chiral PT alkylation.

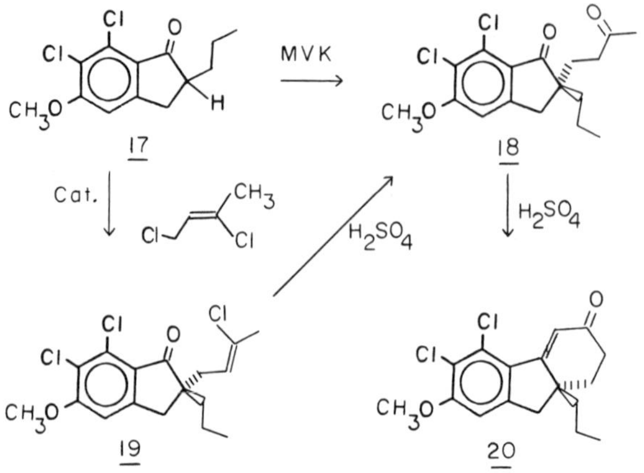

Figure 14. Reaction scheme for the chiral Robinson annulation using the Wichterle reagent.

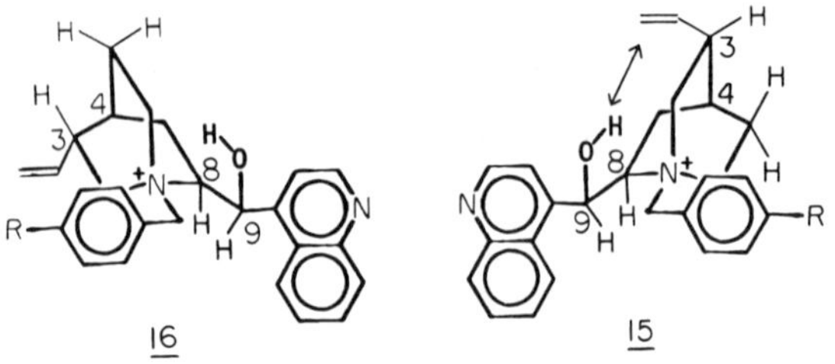

Figure 15. Interactions between the C_9-OH group and the C_3-vinyl group in chinchonine (15) and cinchonidine (16) derived catalysts.

Literature Cited

1. Preliminary Communications: Dolling, U.-H.; Davis, P.; Grabowski, E.J.J. J. Am. Chem. Soc. 1984, 106, 446. Hughes, D.L.; Dolling, U.-H.; Ryan, K.M.; Grabowski, E.J.J.; Schoenewaldt, E.F. Abstracts, 188th ACS National Meeting in Philadelphia, Pennsylvania, August 26-30, 1984, No. 225, Div. of Org. Chem.
2. Duggan, P.G.; Murphy, W.S. J. Chem.Soc., Perkin I 1976, 634.
3. Duhamel, P.; Valnot, J.-Y.; Jamal Eddine, J. Tetrahedron Lett. 1983, 2863.
4. Pfau, M.; Revial, G.; Guingant, A.; d'Angelo, J. J. Am. Chem. Soc. 1985, 107, 273.
5. Enders, D. Chemtech 1981, 504.
6. Meyers, A.I.; Williams, D.R.; Erickson, G.W.; White, S.; Druelinger, M. J. Am. Chem. Soc. 1981, 103, 3081.
7. Tomioka, K.; Ando, K.; Takemasa, Y.; Koga, K. J. Am. Chem. Soc. 1984, 106, 2718.
8. Hashimoto, S.; Koga, K. Tetrahedron Lett. 1978, 573.
9. Enders, D.; Eichenauer, H. Chem. Ber. 1979, 112, 2933.
10. Fiaud, J.-C. Tetrahedron Lett. 1975, 3495.
11. Saigo, K.; Koda, H.; Nohita, H. Bull. Chem. Soc. Jpn. 1979, 52, 3119.
12. Julia, S.; Ginebreda, A.; Guixer, J.; Thomas, A. Tetrahedron Lett. 1980, 3709.
13. Pluim, H.; Wynberg, H. J. Org. Chem. 1980, 45, 2498.
14. Harigaya, Y.; Yamagucchi, H.; Onda, M. Heterocycles 1981, 15, 183.
15. Cram, D.J.; Sogah, G.D.Y. J. Chem. Soc., Chem. Commun. 1981, 625.
16. Hermann, K.; Wynberg, H. J. Org. Chem. 1979, 44, 2238.
17. Colonna, S.; Re, A.; Wynberg, H. J. Chem. Soc. Perkin I 1981, 547.
18. Wynberg, H. Recl. Trav. Chem. Pays.-Bas 1981, 100, 393.
19. Wynberg, H. Chemtech 1982, 116.
20. de Solms, S.J.; Woltersdorf, O.W., Jr.; Cragoe, E.J., Jr. J. Med. Chem. 1978, 21, 437.
21. Woltersdorf, O.W., Jr. J. Labelled Compd Radiopharm. 1980, 17, 635.
22. Irvin, J.D.; Vlasses, P.H.; Huber, P.B.; Feinberg, J.A.; Ferguson, R.K.; Scrogie, J.J.; Davies, R.O. Clin. Pharmacol. Ther. 1980, 27, 260.
23. Zacchei, A.G.; Dobrinska, M.R.; Wishowsky, T.I.; Kwan, K.C.; White, S.D. Drug Metab. Dispos. 1982, 10, 20.
24. Dehmlow, E.V.; Heider, J. J. Chem. Res. Synop. 1981, 292.
25. House, H.O.; Auerback, R.A.; Gall, M.; Peet, N.P. J. Org. Chem. 1973, 38, 514.
26. Cragoe, E.J., Jr.; Stokker, G.E.; Gould, N.P. U.S. Patent 4316043, 1982.
27. Wichterle, O.; Prochaska, J.; Hoffman, J. Coll. Czech. Chem. Commun. 1948, 13, 300.
28. Jung, M.E. Tetrahedron Lett. 1976, 32, 3.
29. Gawley, R.E. Synthesis 1976, 777.

RECEIVED July 26, 1986

Chapter 8

The Phase-Transfer-Assisted Permanganate Oxidation of Alkenes and Alkynes

Donald G. Lee, Eric J. Lee, and Keith C. Brown

Department of Chemistry, University of Regina, Regina, SK, Canada S4S 0A2

Permanganate ion can be easily
solubilized in nonaqueous solvents by the
use of a variety of phase transfer agents
including quaternary ammonium and
phosphonium ions, as well as both cyclic
and acyclic polyethers. Alkenes are
oxidized by permanganate under nonaqueous
conditions to yield either cis-diols or
cleavage products. Alkynes are oxidized
to α-diketones in excellent yields, with
very little cleavage of the carbon-
carbon triple bond. Mechanism studies
have shown that both of these reactions
are initiated by formation of a π-bond
between the double or triple bond and
manganese. The π-complex then
rearranges into a metallocyclooxetane
and subsequently into a cyclic
manganate(V) diester that decomposes to
give the observed products.

The versatility of permanganate as an oxidant has been greatly enhanced in the past decade by the observation that it can be solubilized in nonaqueous solvents with the aid of phase transfer agents (1). The literature contains descriptions for the use of this procedure for the oxidation of alkenes (2-13), alkynes (13-18), aldehydes (19), alcohols (20), phenols (21,22), ethers (23), sulfides (24,25), and amines (20,26). The dehydrogenation of triazolines has also been achieved by the use of permanganate and a phase transfer agent (27).

The Phase Transfer Agents

It has been demonstrated that permanganate can be easily transfered into an organic solvent from either an aqueous solution or from the solid phase. The agents that have been used for these purposes include quaternary ammonium and phosphonium ions (28), crown ethers (4,5,29,30), and linear polyethers (5,30,31). In addition, procedures have been described in which quaternary ammonium and phosphonium permanganates are preformed and isolated as semistable solids that can then be used as general oxidants in a wide variety of solvents (32,33). Precautions should be taken to avoid violent thermal decompositions when the latter procedure is used (34-36).

Several typical extraction constants, K_E, for the transfer of permanganate from an aqueous solution into methylene chloride are listed in Table 1.

Table I. Extraction Constants for the Transfer of Permanganate from Water into Methylene Chloride (32)

Quaternary Ammonium Ion	log K_E [a]
Tetraethylammonium	1.64
Tetra-n-propylammonium	3.71
Tetra-n-butylammonium	4.98
Methyltri-n-butylammonium	4.72

[a] $K_E = [QMnO_4]_{CH_2Cl_2} / [Q^+][MnO_4^-]_{H_2O}$

It can be seen from these data that permanganate can easily be transfered from an aqueous phase into methylene chloride solutions. Quaternary phosphonium ions and polyethers (both cyclic and linear) also exhibit large K_E values (30,32).

The solubilities of the quaternary ammonium and phosphonium permanganates, some of which have been summarized in Table II, provide a measure of the ability of these ions to transfer permanganate from the solid phase. It may be noted that the oxidant which exhibits the highest solubility in the greatest variety of solvents is methyltri-n-octylammonium permanganate. The superior ability of this quaternary ammonium ion to promote the solubility of permanganate has been attributed to a "penetration effect" which allows the anion to be located close to the centre of the cation and, thereby, to be partly shielded from the solvent by the organophilic ligands on the cation.

Quaternary ammonium and phosphonium permanganates exist as intimate ion pairs in nonpolar solvents such as methylene chloride and toluene (1). However, in more polar solvents, such as acetone, nmr studies indicate that they are better described as being solvent separated ion pairs (37). In water, these salts separate completely and exist as individual ions.

Table II. Solubilities of Permanganate Salts (M) (32)

Cation	CH$_2$Cl$_2$	CHCl$_3$	CCl$_4$	C$_6$H$_5$CH$_3$
Tetra-n-butylammonium	0.417	a	2.96x10^5	3.44x10^{-4}
Tetra-n-octylammonium	0.713	0.604	5.93x10^{-4}	4.02x10^{-4}
Methyltri-n-butylammonium	1.83	1.14	5.62x10^{-5}	4.23x10^{-4}
Methyltri-n-octylammonium	1.38	1.07	1.60x10^{-2}	0.798
n-Heptyltriphenylphosphonium	1.36	1.28	b	2.02x10^{-4}
Benzyltriphenylphosphonium	0.430	0.093	b	b
18-Crown-6/K$^+$	0.813	0.257	b	1.64x10^{-4}

a Unstable solution. b Insoluble.

The use of long chain linear polyethers as phase transfer agents has also been investigated. These compounds, which are much less expensive to prepare than the corresponding crown ethers, are also good phase transfer agents for permanganate (30,31). In fact, very long chain polyethylene glycols are capable of transfering more than one mole of permanganate per mole of polyether. From the data in Table III it can be seen that although each mole of crown ether is capable of transfering only one mole of permanganate, the linear polyethers are capable of transfering five or more moles of permanganate per mole of polyether.

Crystallographic studies have shown that polyethylene glycols exist in the solid state as distorted helices containing about seven oxygens per loop (38). When these compounds complex a salt in a nonpolar organic solvent, the oxygens fold inward around the cation and the methylenes present a nonpolar surface to the solvent. A minimum of six oxygens are required to fully complex a potassium ion. However, if the cations were situated within the cavity of the helix there would be a certain amount of electrical repulsion which would discourage the location of ions on adjacent loops. Consequently, it may be seen from the data in Table III that as the number of complexed cations increases, the number of oxygens required per cation also increases.

The Oxidation of Alkenes

The nature of the products obtained from the oxidation of alkenes with aqueous permanganate is known to be controlled primarily by the pH of the medium. Under basic conditions, cis-diols are obtained, while under

Table III. The Solvation of $KMnO_4$ in Methylene Chloride Solutions by Polyethers (30).

Polyether	K_R [a]	n [b]
18-Crown-6	0.99±0.01	6
Dibenzo-18-crown-6	0.95±0.02	6
$CH_3(OCH_2CH_2)_{12}OH$ [c]	1.02±0.02	12
$CH_3(OCH_2CH_2)_{16}OH$ [c]	1.40±0.01	12
$CH_3(OCH_2CH_2)_{45}OH$ [c]	2.91±0.15	15
$CH_3(OCH_2CH_2)_{113}OH$ [c]	5.84±0.04	19

[a] Moles of $KMnO_4$ solubilized per mole of polyether present.
[b] Number of polyether oxygens per mole of $KMnO_4$ solubilized.
[c] Average formula.

neutral conditions the products are α-ketols. When the solutions are acidic, cleavage of the carbon-carbon double bond occurs, and in acetic anhydride solutions α-diketones are obtained (39).

Under phase transfer conditions, where the oxidations are carried out in nonpolar solvents, the nature of the product obtained depends on the way in which the reaction is worked up. Normally, the reactions are carried out with stirring in methylene chloride solutions until the reaction is complete, as evidenced by replacement of the purple permanganate color with an opaque brown due to the formation of MnO_2 (40). The products can then be isolated either by filtering and evaporation of the solvent (13) or by treatment with an aqueous solution. If the aqueous solution is basic, cis-diols are obtained; if it is acidic, cleavage of the carbon-carbon double bond occurs and aldehydes are produced (6). In the later case a further reduction of MnO_2 occurs, possibly because it is involved in the oxidative cleavage reaction (40). When the treatment with water is omitted, the formation of both cis-diols and aldehydes has been reported, the cleavage reaction apparently being more likely if the developing carbonyl is conjugated with an aromatic ring (13).

The Oxidation of Alkynes

When alkynes are oxidized by aqueous permanganate the carbon-carbon triple bond is cleaved, as in Equation 1, with the formation of two moles of carboxylic acid. Only in certain atypical cases are α-diones obtained (41). However, the converse is true under phase transfer conditions where the reactions are carried out in nonaqueous solvents (Equation 2).

$$R-C\equiv C-R' \xrightarrow{MnO_4^-/H_2O} RCOOH + R'COOH \quad (1)$$

$$R-C\equiv R-R' \xrightarrow{QMnO_4/CH_2Cl_2} R-CO-CO-R' \quad (2)$$

Although α-diketones are only rarely isolated from aqueous permanganate oxidations there is conclusive proof that they are intermediates in the reaction (16). α-Diketones are resistant to further oxidation under anhydrous conditions and can thus be isolated in good yields from the phase transfer assisted reactions. However, under aqueous conditions oxidative cleavage takes place rapidly, thus preventing the accumulation of these products.

Reaction Mechanisms

The mechanisms of permanganate oxidations have been the subject of a fairly intensive study which has now lasted for almost a century. While many of these studies were carried out in aqueous solutions, much of what was learned is also germane to an understanding of the reactions which occur in phase transfer assisted reactions. Although most of these studies are interrelated they can conveniently be discussed under the following headings: products, substituent effects, isotope effects, and solvent effects, with the latter being of particular importance to the phase transfer assisted reactions.

Products. The work of Wagner (42,43) and Boeseken (44) established that the oxidation of alkenes under alkaline conditions results in a syn-addition of two hydroxyl groups. In order to account for this observation, Wagner suggested that the reaction must proceed by way of an intermediate cyclic manganate(V) diester, 1, as in equation 3.

Additional evidence in favor of this suggestion was obtained when Wiberg (45) found, by use of oxygen-18 labels, that the reaction takes place by transfer of oxygen from the permanganate to the alkene; i.e., if labeled permanganate is used, the oxygen in the product is also labeled.

Further supporting evidence for this type of a mechanism comes from the consideration of an analogous reaction, the oxidation of alkenes by osmium tetroxide, which also produces cis-diols. In this reaction the

intermediate is so stable that it may be isolated and
characterized. Crystallographic studies have confirmed
that it is a cyclic diester dimer (46).

The evidence supporting the suggestion that cyclic
manganate(V) diesters are intermediates in the reaction
between alkenes and permanganate is compelling (47),
and Simandi (48) has also suggested that a similar
intermediate, 2, may occur during the oxidation of
alkynes as in equation 4.

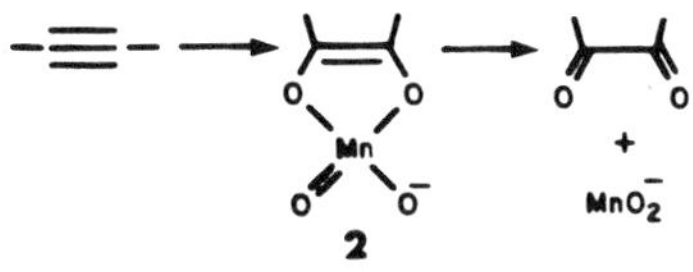

$$-\equiv- \longrightarrow \underset{2}{\overset{\text{Mn}}{\underset{O\diagup\diagdown O^-}{\overset{O\diagdown\diagup O}{\bigvee}}}} \longrightarrow \overset{O}{\underset{}{\bigvee}}\overset{}{\underset{}{\bigvee}}\overset{O}{\underset{}{}} + MnO_2^- \qquad (4)$$

Substituent effects. As the results summarized in
Figure 1a indicate, changing substituents has an almost
negligible effect on the rate of oxidation of
unsaturated compounds by aqueous permanganate.
However, the effects on the rates in nonpolar solvent
systems is dramatically larger and often of apparent
contradiction. For example, the Hammett rho value for
the oxidation of substituted methyl cinnamates and
cinnamic acids by tetrabutylammonium permanganate in
methylene chloride solutions is positive (33,49). See
Figure 1b. However, a rho value of converse sign
(-0.6) is obtained from a Taft plot (Figure 1c) for the
oxidation of vinyl ethers in aqueous tetrahydrofuran
(33,50). For many other compounds the Hammett
relationships are no longer linear, but concave upward;
i.e., the rates of reaction are increased by both
electron donating and electron withdrawing
substituents. This is true, as illustrated in Figure
2, for the oxidation of substituted stilbenes in
aqueous dioxane (33,51), and for the oxidation of both
β-methoxystyrenes and β-bromostyrenes in methylene
chloride (37).

Concave Hammett plots are indicative of a reaction
which undergoes a change in mechanism at the demand of
the substituents (52,53). It may be noted that a
change in the rate limiting step of a multi step
reaction causes the Hammett plots to be concave
downward not upwards as is observed in these examples
(54).

It has been suggested, on the basis of these
curved Hammett plots, that permanganate is ambiphilic,
i.e., capable of acting as both an electrophile and a
nucleophile (55). However, such a suggestion does not
seem reasonable to us. Permanganate, since the charge
is highly delocalized, would not be a very good
nucleophile, and alkenes are not known to be
particularly susceptible to nucleophilic attack.
Instead it is more likely that the reaction is

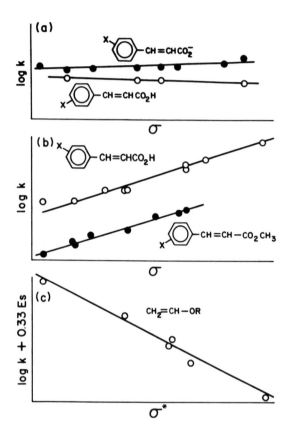

Figure 1. Hammett and Taft plots for the oxidation of unsaturated compounds by permanganate. (a) aqueous solutions, (b) methylene chloride, and (c) aqueous THF.

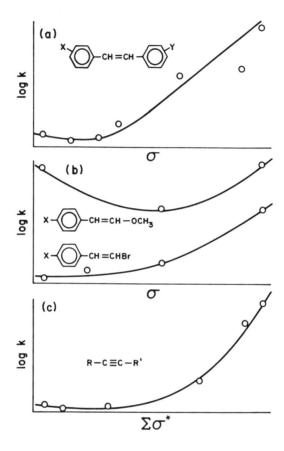

Figure 2. Hammett plots for the oxidation of unsaturated compounds by permanganate in aqueous dioxane (a), and methylene chloride (b) and (c).

initiated, in every instance, by electrophilic attack of manganese on the double bond, as suggested by Sharpless (56). In this way the alkene becomes a ligand on the manganese as in Equation 5. If the double bond then slips (3) it can interact with an oxygen to form the metallocyclooxetane, **4**, which has been predicted theoretically to be an intermediate in these reactions (57).

$$(5)$$

Conversion of **4** into a cyclic manganate(V) diester, which was discussed above, then requires migration of the carbon from manganese to oxygen as in Equation 6.

$$(6)$$

The processes depicted in Equation 5 all involve bond formation, while the reaction in Equation 6 requires cleavage of a carbon-manganese bond and, consequently, it is this step that is most likely to be rate limiting. Since the Hammett plots show that the reaction is accelerated by both electron donating and electron withdrawing substituents, it appears that the reaction can take place via two different transition states, one electron rich and the other electron poor. A possible depiction of this situation is shown in equation 7.

$$(7)$$

Since the mechanism can be altered from one pathway to the other simply by a change in substituents, it follows that the two transition states must be of similar energy and suggests that the reaction can best be visualized by a potential energy diagram. According to transition state theory, the reaction would select the lowest energy pathway from **4** to **1**. When the substrate is capable of accommodating a negative charge, the transition state would resemble **6**, whereas for substrates more capable of bearing a positive charge, the transition state would be more like **5**.

The reaction is less sensitive to changes in substituents when water is used as the solvent, presumably because the more polar solvent can solvate the charged transition states better. Better solvation of these charged species will increase their stability, increase the rates of reaction, and possibly shift the transition state to a point along the reaction coordinate so that it has a greater resemblance to **1**. As will be discussed presently, this suggestion receives some support from a consideration of the observed isotope effects.

The Taft plot for the oxidation of substituted alkynes, as illustrated in Figure 2c, is also concave upward (17). Consequently, it appears as if this reaction may also have a dual mechanism. One possibility has been summarized in Equations 8 and 9.

The electron rich pathway (via **8**) would be followed when electron-withdrawing substituents are present, whereas the electron deficient pathway (via **9**) would be followed when electron donating substituents are present.

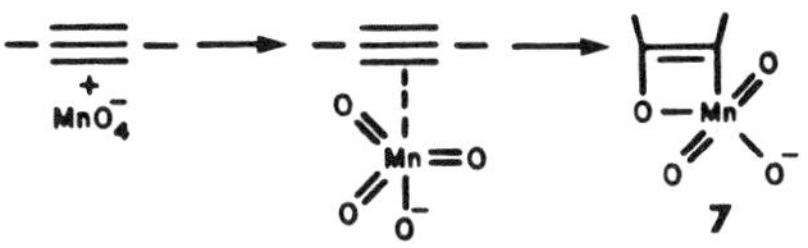

(8)

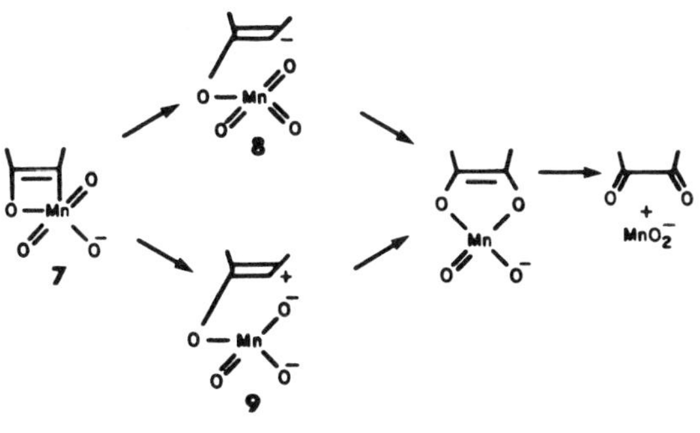

(9)

Isotope effects. The changes in rate caused by
introducing a deuterium atom into the double bond have
been summarized in Table IV.
Inverse isotope effects such as those found in
Table IV are associated with a change in hybridization
at the site of deuterium substitutions (60). Hence, it
is clear from these data that the transition state is
symmetrical with respect to hydridization changes at
both olefinic carbons when the reaction is carried out
in aqueous solutions. However the last four entries in
Table IV indicate that for the oxidation of methyl
cinnamate in methylene chloride solutions only the
β-carbon has undergone a hybridization change (from sp^2

Table IV. Isotope Effects on the Oxidation of Alkenes by Potassium Permanganate

Substrate	Solvent	k_H/k_D	Reference
$C_6H_5CH=CDCO_2H$	0.99M $HClO_4$	0.77	58
$C_6H_5CD=CHCO_2H$	0.99M $HClO_4$	0.75	58
$C_6H_5CH=CDCO_2^-$	0.5M NaOH	0.87	59
$C_6H_5CD=CHCO_2^-$	0.5M NaOH	0.89	59
$C_6H_5CH=CDCO_2^-$	pH 11.3	0.93	59
$C_6H_5CD=CHCO_2^-$	pH 11.3	0.93	59
$C_6H_5CD=CHCO_2CH_3$	CH_2Cl_2	0.91[a]	33
$C_6H_5CH=CDCO_2CH_3$	CH_2Cl_2	1.0[a]	33
$C_6H_5CD=CHCO_2CH_3$	CH_2Cl_2	0.94[b]	33
$C_6H_5CH=CDCO_2CH_3$	CH_2Cl_2	1.0[b]	33

[a] Tetrabutylammonium permanganate used as the oxidant.
[b] (p-Fluorobenzyl)triethylammonium permanganate used as the oxidant.

to sp^3). Consequently, it may be assumed that the
transition state in aqueous solutions may resemble the
cyclic manganate(V) diester intermediate, 1, whereas
under phase transfer conditions the transition state
occurs slightly earlier in the reaction and resembles
the intermediate 6 depicted in Equation 7.

Solvent Effects. Information on the effect of solvent
polarity of the phase transfer assisted permanganate
oxidation of alkenes has been obtained by studying the
oxidation of methyl cinnamate by tetrabutylammonium
permanganate in two different solvents, acetone and
methylene chloride (37).
Three important and interrelated observations were
made: Firstly, it was found that the Hammett rho value
is substantially larger when the solvent is acetone
(1.43), than when it is methylene chloride (0.95).
Secondly, the rates of reaction are faster in methylene
chloride and, thirdly, the rates of reaction in acetone
solutions are independent of the identity of the

quaternary ammonium ion, whereas in methylene chloride solutions significant variations are observed. See Table V.

Taken together these results suggest that the quaternary ammonium ions must not form closely associated ion pairs in acetone solutions. In this more polar solvent, it appears as if the reactants are present as solvent separated ion pairs and that the rate of reaction is, consequently, not effected by the structure of the quaternary ammonium ion. In methylene chloride solutions, where theory predicts tighter ion pairs (61), the ions must be intimately associated in either (or both) the ground state and the transition

Table V. Second Order Rate Constants ($M^{-1}s^{-1}$) for the Oxidation of Methyl Cinnamate by Quaternary Ammonium Permanganates[a]

Quaternary cations	Acetone	Methylene Chloride
Tetra-n-butylammonium	0.32±0.01	1.15±0.02
Tetra-n-octylammonium	0.31±0.01	0.91±0.02
Methyltri-n-octylammonium	0.33±0.01	1.54±0.05
Methyltriphenylphosphonium	0.34±0.02	1.35±0.02
n-Butyltriphenylphosphonium	0.27±0.01	0.99±0.02
Benzyltriphenylphosphonium	0.30±0.01	1.53±0.04
18-Crown-6-K$^+$	0.27±0.01	1.64±0.01

[a] Temp. = 22°. [MnO$_4$]=4x10^{-4} M. [Methyl cinnamate]= 8x10^{-3} M.

state. Furthermore, close contact of the ions seems to decrease the overall polarity of the transition state (as evidenced by the decreased rho value) and thus to increase the rate of reaction. In acetone, where the ion pair is much looser, the effect of substituents is greater because of a greater net charge developed in the transition state.

Acknowledgments

The authors are grateful to the Natural Sciences and Engineering Research Council of Canada, and the Carus Chemical Company for financial support.

Literature Cited

1. Lee, D.G. In "Oxidation in Organic Chemistry, Part D"; Trahanovsky, W.S., Ed.; Academic Press: New York, **1982**; Chapter 2.
2. Starks, C.M. J. Am. Chem. Soc. **1971**, 93, 195.
3. Sala, T.; Sargent, M.V. J. Chem. Soc., Chem. Commun. **1978**, 253.

4. Sam, D.J.; Simmons, H.E. J. Am. Chem. Soc. 1972, 94, 4024.
5. Lee, D.G.; Chang, V.S. J. Org. Chem. 1978, 43, 1532.
6. Ogino, T.; Mochizuki, K. Chem. Lett. 1979, 443.
7. Okimoto, T.; Swern, D. J. Am. Oil Chem. Soc. 1977, 54, 862A.
8. Foglia, T.A.; Barr, P.A.; Malloy, A.J. J. Am. Oil Chem. Soc. 1977, 54, 858A.
9. Weber, W.P.; Sheperd, J.P. Tetrahedron Lett. 1972, 4907.
10. Lampman, G.M.; Sharpe, S.D. J. Chem. Ed. 1983, 60, 503.
11. Marmor, S. "Laboratory Methods in Organic Chemistry"; Burgess: Minneapolis, 1981, p.352-356.
12. Harris, J.M.; Case, M.G. J. Org. Chem. 1983, 48, 5390.
13. Bhushan, V.; Rathore, R.; Chandrasekaran, S. Synthesis, 1984, 431.
14. Krapcho, A.P.; Larson, J.R.; Eldridge, J.M. J. Org. Chem. 1977, 42, 3749.
15. Lee, D.G.; Lamb, S.E.; Chang, V.S. Org. Synth. 1981, 60, 11.
16. Lee, D.G.; Chang, V.S. J. Org. Chem. 1979, 44, 2726.
17. Lee, D.G.; Lee, E.J.; Chandler, W.D. J. Org. Chem. 1985, 50, 4306.
18. Lee, D.G.; Chang, V.S. Synthesis 1978, 462.
19. Menger, F.M.; Rhee, J.U.; Rhee, H.K. J. Org. Chem. 1975, 40, 3803.
20. Schmidt, H.J.; Schafer, H.J. Angew. Chem, Int. Ed. Engl. 1981, 20, 104, 109.
21. Gokel, G.W.; Durst, H.D. Synthesis 1976, 168.
22. Bock, H.; Jaculi, D. Angew. Chem., Int. Ed. Engl. 1984, 23, 305.
23. Schmidt, H.J.; Schafer, H.J. Angew. Chem., Int. Ed. Engl. 1979, 18, 68, 69.
24. Scholz, D. Monatsh. Chem. 1981, 112, 241.
25. Lee, D.G.; Srinivasan, N.S. Sulfur Lett. 1981, 1, 1.
26. Rossi, L.M.; Trimarco, P. Synthesis, 1978, 743.
27. Kadaba, P.K. J. Prak. Chem. 1982, 324, 857.
28. Starks, C.M.; Liotta, C. "Phase Transfer Catalysis"; Academic Press: New York, 1978.
29. Pederson, C.J. J. Am. Chem. Soc. 1967, 89, 7017.
30. Lee, D.G.; Karaman, H. Can. J. Chem. 1982, 60, 2456.
31. Harris, M.J.; Case, M.G.; J. Org. Chem. 1983, 48, 5390.
32. Karaman, H.; Barton, R.J.; Robertson, B.E.; Lee, D.G. J. Org. Chem. 1984, 49, 4509.
33. Lee, D.G.; Brown, K.C. J. Am. Chem. Soc. 1982, 104, 5076.
34. Morris, J.A.; Mills, D.C. Chem. Ind. (London) 1978, 446.

35. Jager, H.; Lutolf, J.; Meyer, M.W. <u>Angew. Chem., Int. Ed. Engl.</u> **1979**, 18, 786.
36. Schmidt, H.J.; Schafer, H.J. <u>Angew. Chem., Int. Ed. Engl.</u> **1979**, 18, 787.
37. Lee, D.G.; Brown, K.C.; Karaman, H. <u>Can. J. Chem.</u> In press.
38. Takahashi, Y.; Tadokoro, H. <u>Macromolecules</u> **1973**, 6, 672.
39. Arndt, D. "Manganese Compounds as Oxidizing Agents in Organic Chemistry"; Open Court:La Salle, Illinois, **1981**; Chapter 5.
40. Perez-Benito, J.F.; Lee, D.G. <u>Can. J. Chem.</u> **1985**, 63, 3545.
41. Khan, N.A.; Newman, M.S. <u>J. Org. Chem.</u> **1952**, 17, 1063.
42. Wagner, G. <u>Chem. Ber.</u> **1888**, 21, 1230, 3347.
43. Wagner, G. <u>Zh. Russ. Fiz-Khim. Obstichest.</u> **1895**, 27, 219.
44. Boeseken, J. <u>Rec. trav. chim.</u> **1921**, 40, 553.
45. Wiberg, K.B.; Saegebarth, K.A. <u>J. Am. Chem. Soc.</u> **1957**, 79, 2822.
46. Schroder, M. <u>Chem. Rev.</u> **1980**, 80, 187.
47. Stewart, R. In "Oxidation in Organic Chemistry, Part A"; Wiberg, K.B., Ed.; Academic Press: New York, 1965; Chapter 1.
48. Simandi, L.I.; Jaky, M. <u>J. Chem. Soc. Perkin Trans. 2</u>, **1977**, 630.
49. Perez-Benito, J.F. Unpublished.
50. Toyoshima, K.; Okuyama, T.; Fueno, T. <u>J. Org. Chem.</u> **1980**, 45, 1600.
51. Henbest, H.B.; Jackson, W.R.; Robb, B.C.G. <u>J. Chem. Soc. B</u> **1966**, 803.
52. Leffler, J.E.; Grunwald, E. "Rates and Equilibria of Organic Reactions". Wiley, New York: **1963**; p.187-191.
53. Exner, O. In "Advances in Linear Free Energy Relationships". Chapman, N.B. and Shorter, J. Ed.; Plenum: New York, **1972**; p.12-17.
54. Jencks, W.P. "Catalysis in Chemistry and Enzymology". McGraw-Hill: New York, **1969**, p.480-487.
55. Freeman, F.; Kappos, J.C. <u>J. Am. Chem. Soc.</u> **1985**, 107, 6628.
56. Sharpless, K.B.; Teranishi, A.Y.; Backvall, J.E. <u>J. Am. Chem. Soc.</u> **1977**, 99, 3120.
57. Rappe, A.K.; Goddard, W.A., III; <u>J. Am. Chem. Soc.</u> **1982**, 104, 3287.
58. Lee, D.G.; Brownridge, J.R. <u>J. Am. Chem. Soc.</u> **1974**, 96, 5517.
59. Lee, D.G.; Nagarajan, K. <u>Can. J. Chem.</u> **1985**, 63, 1018.
60. Halevi, E.A. <u>Progr. Phys. Org. Chem.</u> **1963**, 1, 109.
61. Brandstrom, A. <u>Adv. Phys. Org. Chem.</u> **1977**, 15, 267.

RECEIVED July 26, 1986

Chapter 9

New Developments in Polymer Synthesis by Phase-Transfer Catalysis

Virgil Percec

Department of Macromolecular Science, Case Western Reserve University, Cleveland, OH 44106

> Some particularities of the extraction of ions from an aqueous organic phase, and of the phase catalyzed polyetherification will be summarized. These will represent the fundamentals of our work on the synthesis of some novel classes of functional polymers and sequential copolymers. Examples will be provided for the synthesis of: functional polymers containing only cyclic imino ethers or both cyclic imino ethers as well as their own cationic initiator attached to the same polymer backbone; ABA triblock copolymers and (AB)n alternating block copolymers; and a novel class of main chain thermotropic liquid crystalline polymers containing functional chain ends, i.e., polyethers.

Phase transfer catalysis, a term which has been coined by Starks in 1971(<u>1</u>), became, within only a short period of time, an active subject of research with deep implications especially in preparative organic, organometallic and polymer chemistry(<u>2-7</u>). Traditional fields of polymer chemistry like radical, anionic and condensation polymerizations, as well as chemical modification of polymers, have substantially benefited from the use of phase transfer catalysis. Some of the most significant progress made in this field by exploiting the phase transfer catalysis concept has been the subject of a previous ACS meeting, and it's proceedings were published in a recent book(<u>7</u>). Our research group has become active in this field only recently, and it is the aim of this paper to shortly review some of the work accomplished, or in progress within our laboratory.

Extraction of Ions from an Aqueous Solution. Its Implications on the Mechanism of Phase Transfer Catalyzed Polyetherifications

For a two phase system (water and a water nonmiscible organic solvent) containing a hydrophobic salt QX dissolved in water $(Q_W^+ + X_W^- \rightleftharpoons QX_S)$, the conditional extraction constant e* is defined as $C_{QXs} = e_{QX}[Q^+]_W [X^-]_W$, where C_{QXs} is the total concentration of Q (quat) and X^- (corresponding anion) present in the organic phase in the molar ratio 1 to 1; and $[Q^+]_W$ and $[X^-]_W$ are the respective

0097-6156/87/0326-0096$06.00/0
© 1987 American Chemical Society

concentrations in the water phase. Since the Q part of the QX salt is always hydrophobic, it is the entropic factor which will tend to drive the QX compound out of the aqueous layer. Therefore, the extraction constants are strongly dependent on the structure of both the anion (X⁻) and cation (Q⁺) (or in other words on their hydrophobicity) and on the solvent nature, but due to the dominant influence of entropy factor, it is not very dependent on the temperature. For a specific Q⁺, the value of the e* will be controlled by the hydrophobicity of the X⁻, and there are quantitative examples in the literature demonstrating this(3). It is known for example that the extraction constant of a salt having PhO⁻ as anion is higher by a factor of about 10^5 than the extraction constant of a salt having OH⁻ as anion. Therefore, it can be certainly assumed that when the PhO⁻ anion becomes a polymer chain end, the extraction constant of its salt with Q⁺ should be molecular weight dependent based on the fact that the hydrophobicity of the phenoxy anion increases with the increase of the polymers molecular weight, i.e., e*$_{PhO-(PhO)n-PhO-Q^+}$ >> e*$_{PhO-Q^+}$. This conclusion has some important implications for the two-phase phase transfer catalyzed reactions especially when the transfer of the ion-pair into the organic phase is the rate determining step. Under these conditions we can speculate that the nucleophilicity of a phenol attached to a polymer chain end is molecular weight dependent and increases with the increase of the polymer molecular weight. This behavior can be considered as a unique situation in which the reactivity of a functional group attached to a polymer chain end increases with the increase of the polymer molecular weight. This particularity, permits us to tailor quantitative reactions of ω-phenol oligomers and α,ω-bisphenol oligomers with both electrophilic low molecular weight compounds and with α,ω-di(electrophilic) oligomers(8-14). An important side reaction which has to be avoided in both cases is the displacement of the electrophile present in the organic phase with the OH⁻ transferred from the water phase into the organic phase. When working with α,ω-bisphenol oligomers, this side reaction can be avoided by using solvents which will usually provide a low extraction constant for the transfer of the anions from water to organic phase. Since polymeric onium phenolates have very large extraction constants, the use of aromatic solvents will depress their extraction into the organic phase to the level of a low molecular weight phenol, but at the same time the transfer of the OH⁻ into organic phase will be decreased below the level it can compete with the other nucleophile. We have demonstrated that this situation can be easily accomplished(15).

A short comparative discussion of conventional and phase transfer catalyzed step polymerizations would let us point out some basic differences between these two reactions. Conventional step polymerization is a statistical reaction whose kinetic treatment is based on the equal reactivity of functional groups participating in polymerization, indifferent of the molecular weight of the polymer at which chain ends are attached. For an equimolar ratio of the two monomers, the polymerization degree can be calculated from the "extent of reaction" p, i.e., $\overline{DP}$ = 1/(1-p). Only at very high conversions (higher than 99.5%) and only when stoichiometric monomer ratios are used (i.e., 1:1) can high degrees of polymerization be obtained. For a 100 percent conversion, the theoretical polydispersity of the obtained polymer is equal 2, i.e., $\overline{M}w/\overline{M}n = 2.0$. When the reaction is performed with a 1:1 mole ratio of the two monomers, at any time including at 100 percent conversion the polymer chain ends will have a statistical distribution of the functional groups attached to them.

Several particularities of phase transfer catalyzed polyetherification are as follows. Stoichiometric phase transfer catalyzed polymerizations do not take place between stoichiometric ratio of monomers, since the nucleophilic monomer is always transferred in a small amount into the organic phase. Consequently, because their reaction is a non-stoichiometric one there is no need for an equimolar ratio between the two monomers to get polymers with high molecular weights. High molecular weight polymers are usually obtained also at low conversions. In several cases, even at 100 percent conversion the polydispersity of the obtained polymers is low, i.e., $\overline{M}w/\overline{M}n \leq 1.3$. At any conversion, the organic phase contains only polymers with electrophilic chain ends, even when the nucleophilic monomer was used in excess.

At this time, only some of these particularities can receive an explanation. Onium bisphenolates are nonsolvated ion-pairs with reduced cation-anion interaction energy, and consequently are very reactive. Their low concentration in the organic phase easily explains the electrophilic nature of the chain ends, at least when the rate determining step is their transfer from the water into the organic phase:

E-E + ⁻N-N⁻ + E-E- ⟶   E-E-N-N-E-E
                        E-E-N-N-E-E-N-N-E-E
                        etc.

The side reaction previously discussed, i.e., the displacement of the electrophilic chain ends by OH⁻, can give rise to a new nucleophile with a different reactivity. This reaction can be avoided, but its presence or absence apparently cannot explain any of the above particularities. A major point of concern refers to the reactivity of nucleophilic and electrophilic groups present in the reaction mixture at different stages of the polymerization. An obvious question would be: is the reactivity of the nucleophilic and electrophilic polymeric chain ends the same with that of the monomeric ones? Apparently, as a result of the previous discussion, the nucleophilicity depends upon and increases with the molecular weight. At this time we cannot say much about the reactivity of the electrophilic groups, although in several cases we have speculated a possible enhanced electrophilicity based on anchimeric assistance([16-17](#)). A final comment concerns the overall polymerization behavior at different stages of the reaction. Usually, the initial concentration of the phase transfer catalyst, represents 5-10 mole % from the nucleophilic monomer, and, therefore, the reaction proceeds under two phase conditions. At high conversion, the concentration of the nucleophile decreases and the amount of phase transfer catalyst reaches and even exceeds 100 mole % from the concentration of the nucleophile. Consequently, at this stage of the reaction the entire amount of nucleophile will be present in solution, and the polymerization can be considered as employing the organic phase only. In other words, even if we have a two-phase system, the water phase no longer plays the same role as before and the reaction can be considered as being a one-phase reaction.

All these particularities of phase transfer catalyzed polyetherification were staying behind our approach to the design of new macromolecules.

Three major topics of research which are based on phase transfer catalyzed reactions will be presented with examples. These refer to the synthesis of functional polymers containing functional groups (i.e., cyclic imino ethers) sensitive both to electrophilic and nucleophilic reagents; a novel method for the preparation of regular, segmented, ABA triblock and (A-B)n alternating block copolymers, and the development of a novel class of main chain thermotropic liquid-crystalline polymers, i.e., polyethers.

Functional Polymers Containing Cyclic Imino Ethers

Scheme 1 presents both the synthesis and the ring opening reactions of 2-(p-hydroxyphenyl)-2-oxazoline (HPO). HPO reacts with NaOH on heating to provide N-(2-hydroxyethyl)-p-hydroxybenzamide, and with weak electrophilic compounds like benzyl bromide or allyl chloride to provide poly [N-(p-hydroxybenzoyl) ethylenimine]. Consequently, the etherification reaction of an α,ω-di(electrophilic) oligomer with the sodium salt of 2-(p-hydroxyphenyl)-2-oxazoline in an aprotic dipolar solvent like DMSO or DMF would be accompanied by these two side reactions, even if the nucleophilicity of the phenolate is higher than that of the oxazoline ring. A phase-transfer-catalyzed Williamson etherification of an α,ω-di(electrophilic) oligomer performed in chlorobenzene-aqueous NaOH and stoichiometric amount of phase transfer catalyst as detailed elsewhere(18,20) gives rise to perfectly bifunctional α,ω-di[2-(p-phenoxy)-2-oxazoline] oligomers. The same bifunctional oligomers can be prepared through a chain extension of an α,ω,-di(phenol) oligomer or bisphenol monomer with methylene chloride as outlined in Scheme 2. The mechanistic reasons for this last successful reaction are outlined in Scheme 3, while Scheme 4 presents the synthetic routes for the preparation of α,ω-di(electrophilic) aromatic polyether sulfones. The detailed synthesis of these last oligomers has been already described elsewhere(15). Figure 1 gives an example of 200 MHz ^{1}H-NMR spectra of the starting α,ω-di(electrophilic) oligomer and the obtained α,ω-di[2-(p-phenoxy)-2-oxazoline] oligomer to demonstrate the quantitative nature of this reaction. Additional examples can be obtained from other previous publications from our laboratory(18,20).

The second case refers to the synthesis of the first example of a bifunctional polymer containing not only cationically polymerizable heterocycles, but also their own cationic initiator as pendant groups(19). Scheme 5 presents the synthesis of a poly(2,6-dimethyl-1,4-phenylene oxide) containing both 2-(p-phenoxy)-2-oxazoline and bromobenzylic pendant groups. Incomplete etherification of a poly(2,6-dimethyl-1,4-phenylene oxide) containing 0.14 -CH$_2$Br groups per structural unit with HPO leads to a polymer containing 0.106 2-(p-phenoxy)-2-oxazoline groups and 0.033 -CH$_2$Br groups per structural unit, corresponding to a 3.21/1 mole ratio between the heterocyclic monomer and its initiator. A 200 MHz ^{1}H-NMR spectrum of this polymer is presented in Figure 2. This polymer is stable at

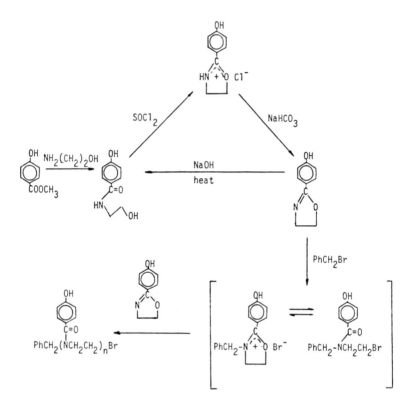

Scheme 1. Synthesis and reactions of 2-(p-hydroxyphenyl)-2-oxazoline.

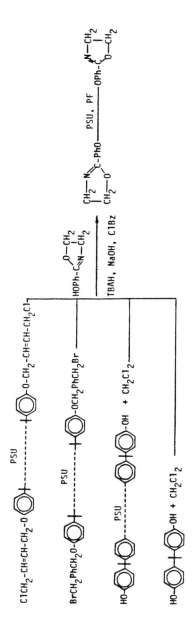

Scheme 2. Synthesis of α,ω-di[2-(p-phenoxy)-2-oxazoline] oligomers.

$$-\text{PhO}^- + \text{CH}_2\text{Cl}_2 \longrightarrow -\text{PhOCH}_2\text{Cl}$$

$$-\text{PhOCH}_2\text{Cl} + \text{OH}^- \longrightarrow -\text{PhO}^- + \text{CH}_2\text{O}$$

$$-\text{PhOCH}_2\text{Cl} + {}^-\text{PhO} \longrightarrow -\text{PhOCH}_2\text{OPh}-$$

$$-\text{PhOCH}_2\text{Cl} + {}^-\text{OPhOxz} \longrightarrow -\text{PhOCH}_2\text{OPhOxz}$$

Scheme 3. Reactions occurring during the preparation of α,ω-di[2-(p-phenoxy)-2-oxazoline] oligomers according to the second and third synthetic routes in Scheme 2.

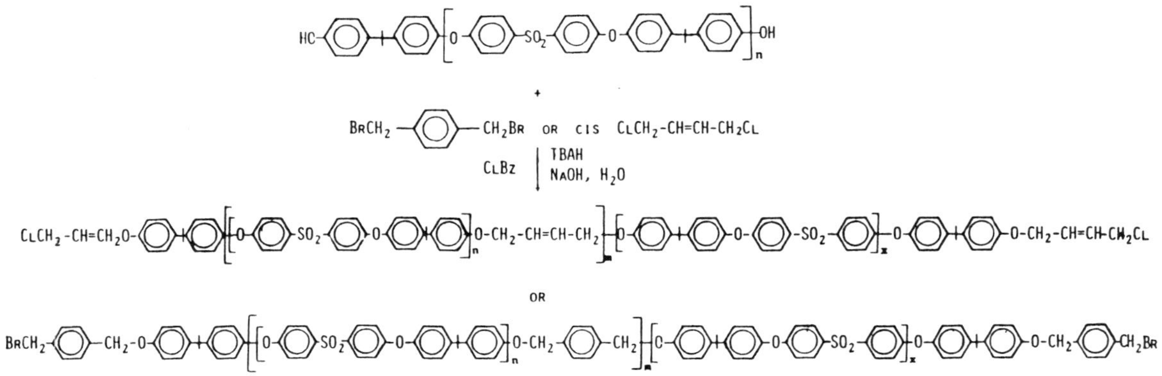

9. PERCEC *Polymer Synthesis by Phase-Transfer Catalysis* 103

Scheme 4. Synthesis of α,ω-di(electrophilic) aromatic polyether sulfones.

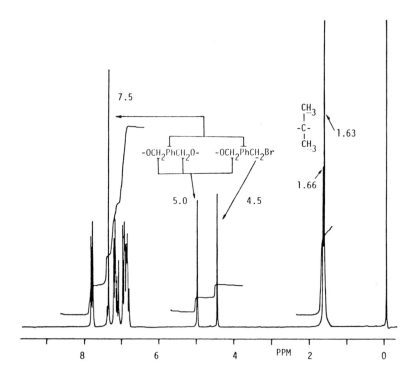

Figure 1a. 200 MHz ^{1}H-NMR spectrum (CDCl$_3$, TMS) of α,ω-di(bromobenzyl aromatic polyether sulfone (PSU) from Scheme 4.

9. PERCEC *Polymer Synthesis by Phase-Transfer Catalysis* 105

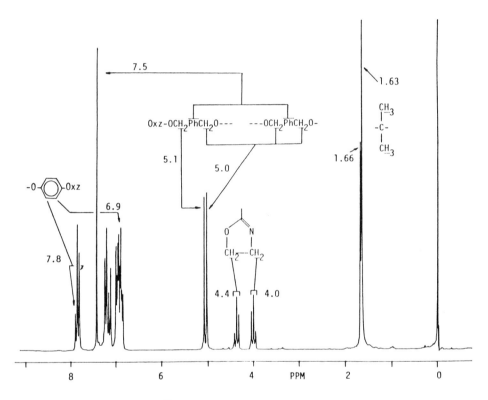

Figure 1b. 200 MHz ^{1}H-NMR spectrum (CDCl$_3$, TMS) of α,ω-di[2-(p-phenoxy)-2-oxazoline] aromatic polyether sulfone obtained from α,ω-di(bromobenzyl) PSU.

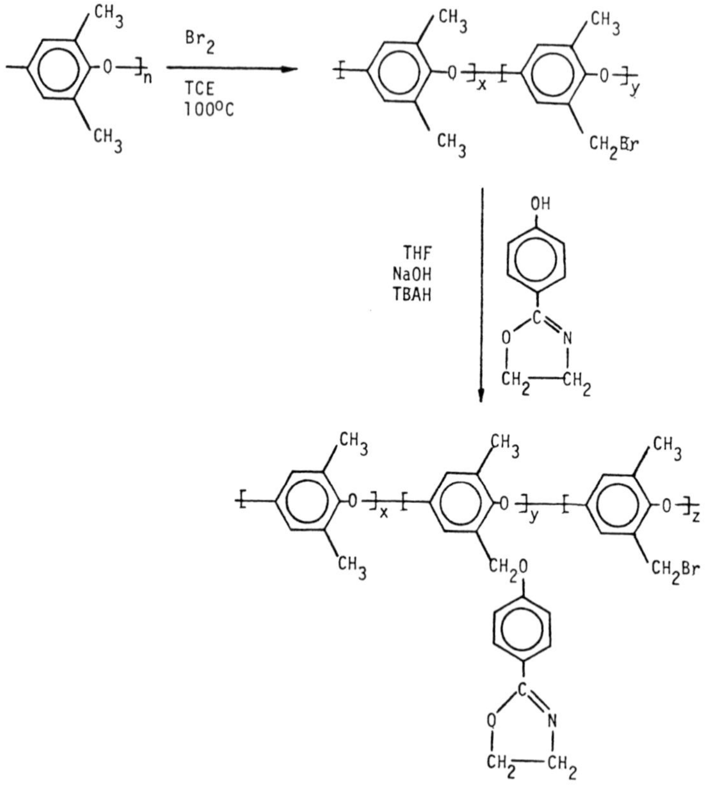

Scheme 5. Synthetic routes used for the preparation of functional polymers containing pendant 2-oxazoline and bromobenzyl groups.

room temperature, but upon heating above its glass transition temperature, the -CH$_2$Br groups initiate the cationic ring opening polymerization of the cyclic imino ether giving rise to a polymer network and a substantial increase in the glass transition temperature. This is illustrated by the Figure 3.

Phase Transfer Catalyzed Polyetherification
of Chain Ended Functional Polymers, a
New Method for the Synthesis of Sequential Copolymers

We have recently demonstrated that the phase transfer catalyzed polyetherification of an α,ω-di(electrophilic) or α,ω-di(nucleophilic) oligomer with a bisphenol or an α,ω-di(electrophilic) oligomer represents a new and very efficient method for the synthesis of regular copolymers(8,21). The chain extension of two different α,ω-di(phenol) oligomers with a dielectrophilic monomer again through a phase transfer catalyzed etherification gives rise to segmented copolymers(22).
Under carefully selected reaction conditions, the polyetherification of an ω-phenol oligomer with an α,ω-di(electrophilic) oligomer produces unexpectedly pure ABA triblock copolymers(11), while the polyetherification of an α,ω-di(electrophilic) oligomer with an α,ω-di(nucleophilic) oligomer represents a new method for the synthesis of perfectly alternating (AB)$_n$ block copolymers(9-12,22).
An example for the synthesis of poly(2,6-dimethyl-1,4-phenylene oxide) - aromatic poly(ether-sulfone) - poly(2,6-dimethyl-1,4-phenylene oxide) ABA triblock copolymer is presented in Scheme 6. Quantitative etherification of the two polymer chain ends has been accomplished under mild reaction conditions detailed elsewhere(11). Figure 4 presents the 200 MHz ^{1}H-NMR spectra of the ω-(2,6-dimethylphenol) poly(2,6-dimethyl-1,4-phenylene oxide), of the α,ω-di(chloroally) aromatic polyether sulfone and of the obtained ABA triblock copolymers as convincing evidence for the quantitative reaction of the parent polymers chain ends. Additional evidence for the very clean synthetic procedure comes from the gel permeation chromatograms of the two starting oligomers and of the obtained ABA triblock copolymer presented in Figure 5.

Thermotropic Polyethers -
A New Class of Main Chain Liquid Crystalline Polymers

Recently we have developed a new class of thermotropic liquid crystalline (LC) main-chain polymers, i.e., polyethers of mesogenic bis-phenols(16-17,23-26). Since the obtained polymers are not soluble in dipolar aprotic solvents, the only available synthetic avenue for their preparation consists in the phase transfer catalyzed polyetherification.
There are several very important advantages we obtain from this novel class of LC polymers. They can be prepared with well defined chain ends and narrow molecular weight distribution. Therefore, we could provide important information concerning the influence of the nature of the polymer chain ends on their mesomorphic behavior(16,23-24). They are soluble in conventional solvents, have lower melting and isotropization temperatures and still broader thermal stability of the mesophase than the corresponding poly-

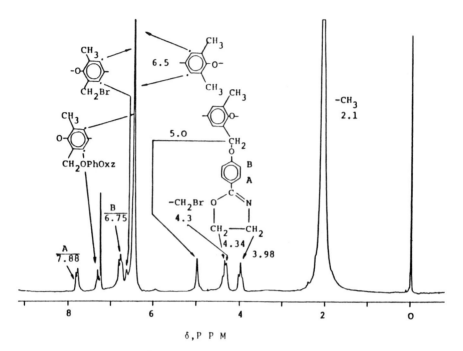

Figure 2. 200 MHz ^{1}H-NMR spectrum (CDCl$_3$) of the poly(2,6-dimethyl-1,4-phenylene oxide) containing 2-oxazoline and bromobenzyl pendant groups.

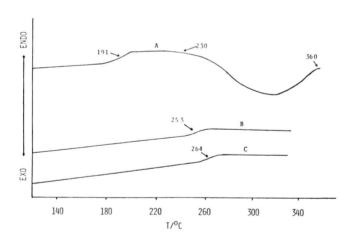

Figure 3. DSC heating scans (20°C/min) of the poly(2,6-dimethyl-2,4-phenylene oxide) containing 2-oxazoline and bromobenzyl pendant groups. A) second heating scan (first heating scan up to 200°C); B) third heating scan; C) fourth heating scan (after annealing 30 min. at 270°C).

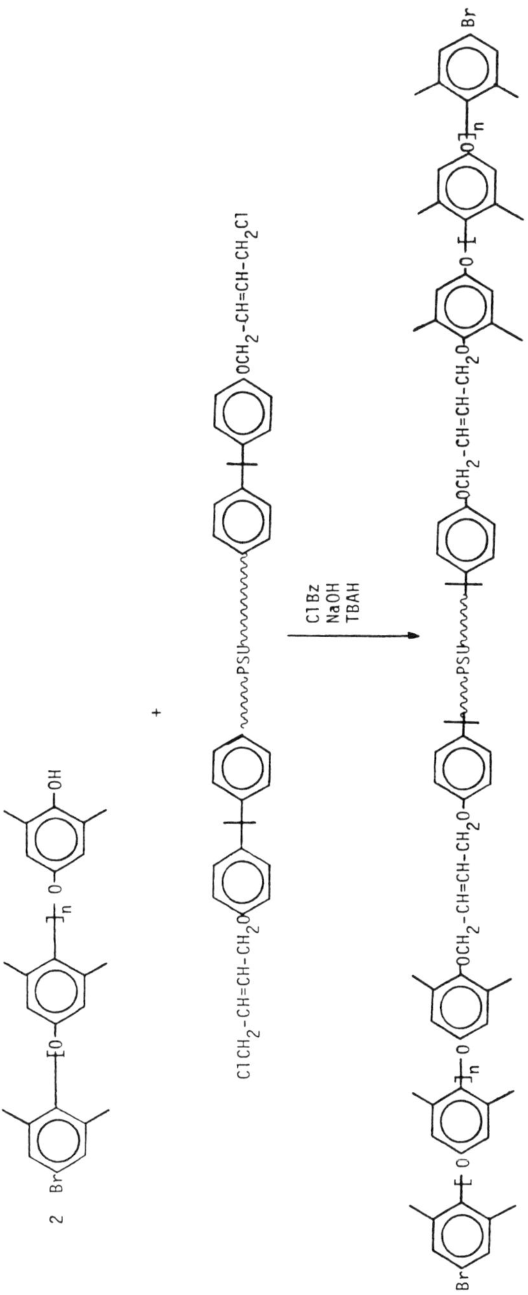

Scheme 6. Synthesis of poly(2,6-dimethyl-1,4-phenylene oxide)-aromatic polyether sulfone-poly(2,6-dimethyl-1,4-phenylene oxide) (PPO-PSU-PPO) triblock copolymer.

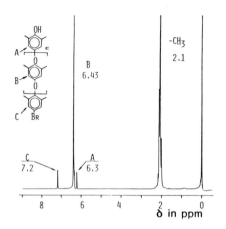

Figure 4a. 200 MHz ^{1}H-NMR spectrum of ω-(2,6-dimethylphenol) poly(2,6-dimethyl-1,4-phenylene oxide) (PPO) ($\overline{M}_n$ = 2,235, CCl$_4$, TMS).

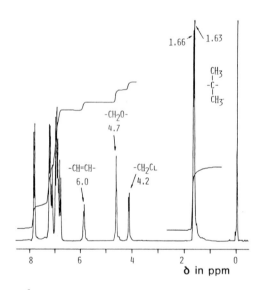

Figure 4b. 200 MHz ^{1}H-NMR spectrum of α,ω-di(chloro allyl) aromatic polyether sulfone (PSU) from Scheme 4 ($\overline{M}_n$ = 1,930, CDCl$_3$, TMS).

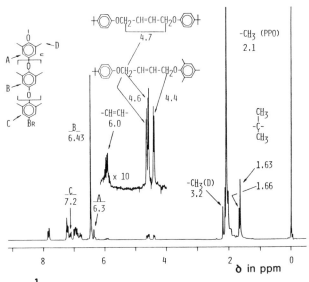

Figure 4c. 200 MHz ^{1}H-NMR spectrum of the PPO-PS-PPO block copolymer (CDCl$_3$, TMS).

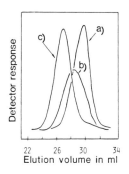

Figure 5. GPC curves of: A) ω-(2,6-dimethylphenol) PPO, (M_n = 1,930); B) α,ω-di(chloroallyl) PSU ($\overline{M}_n$ = 3,900); and C) PPO-PSU-PPO triblock copolymer.

esters(25). The synthesis of LC copolyethers allows us to tailor their thermal transitions over a very broad range of temperatures(23-25). At the same time, they let us obtain information concerning the influence of sequence distribution on their mesomorphic properties(26). Last but not least, by varying the previously discussed methods for the synthesis of sequential copolymers we could develop the first classes of both segmented copolymers and alternating block copolymers containing thermotropic LC segments in one block and either elastomeric or thermoplastic segments as the second block(27). Scheme 7 outlines an example for the synthesis of the thermotropic polyethers and copolyethers based on 4,4'-dihydroxybiphenyl(25). A comparison between the thermal transitions of a set of polyethers and the corresponding polyesters is presented in Figure 6. Figure 7 demonstrates the ability to tailor the thermal stability range of the mesophase through copolyetherification.

There are several other active topics under examination in our laboratory, for example, surface modification of polymers under phase transfer catalyzed reactions and single electron transfer phase transfer catalyzed polymerizations. The limited space, however, precludes discussion here.

What will phase transfer catalysis provide polymer chemistry within the near future? It is apparently still to early to predict this. We are not yet in the possession of many elemental mechanistic and kinetic understandings in order to answer questions like, for example, why not "living polyetherification?"

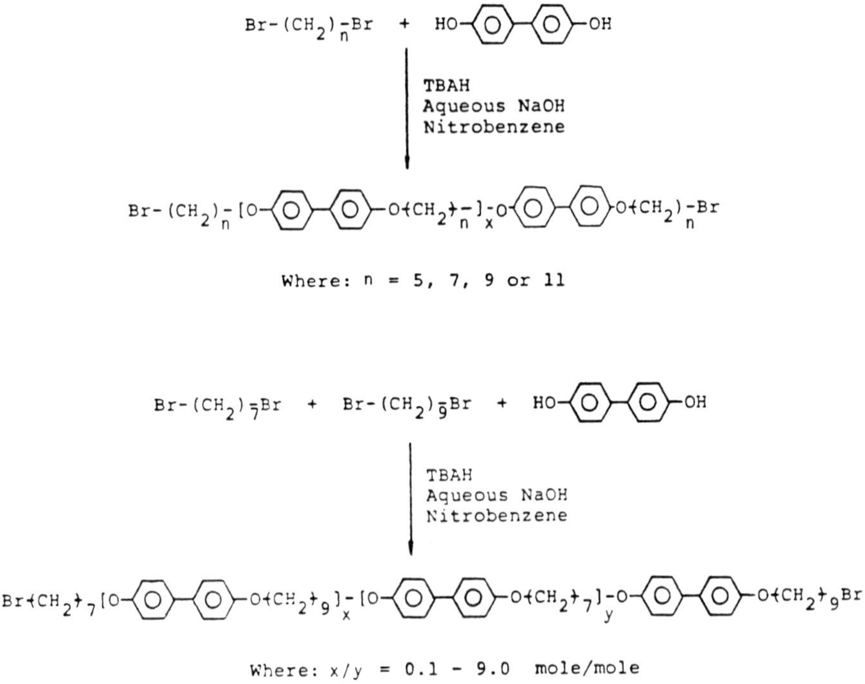

Br-(CH$_2$)$_n$-Br + HO-⟨O⟩-⟨O⟩-OH

TBAH
Aqueous NaOH
Nitrobenzene

Br-(CH$_2$)$_n$-[O-⟨O⟩-⟨O⟩-O(CH$_2$)$_n$]$_x$-O-⟨O⟩-⟨O⟩-O(CH$_2$)$_n$-Br

Where: n = 5, 7, 9 or 11

Br-(CH$_2$)$_7$-Br + Br-(CH$_2$)$_9$-Br + HO-⟨O⟩-⟨O⟩-OH

TBAH
Aqueous NaOH
Nitrobenzene

Br(CH$_2$)$_7$-[O-⟨O⟩-⟨O⟩-O(CH$_2$)$_9$]$_x$-[O-⟨O⟩-⟨O⟩-O(CH$_2$)$_7$]$_y$-O-⟨O⟩-⟨O⟩-O(CH$_2$)$_9$Br

Where: x/y = 0.1 - 9.0 mole/mole

Scheme 7. Synthetic avenues used for the synthesis of 4,4'-dihydroxybiphenyl polyethers and copolyethers.

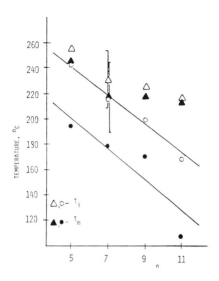

Figure 6. Thermal transition temperatures (T_m = melting, T_i = isotropization) versus n, the number of methylene units in the polymers containing 4,4'-dihydroxybiphenyl for: 1) polyethers (●) T_m, (o) T_i; 2) polyesters (▲) T_m, (△) T_i (data from reference 25).

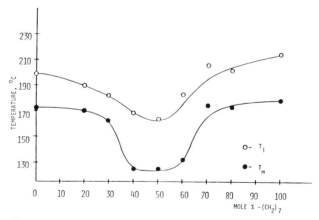

Figure 7. Thermal transition temperatures (T_m, T_i) versus the mole % of 1,7-dibromoheptane in the reaction mixture, for the homo- and copolymers based on 4,4'-dihydroxybiphenyl, 1,7-dibromoheptane, and 1,4-dibromononane.

Acknowledgments

Most of this work has been supported by the Polymers Program of the National Science Foundation under Grant No. DMR 82-13895. I am deeply grateful to my dedicated coworkers listed as coauthors in the referenced papers.

References

1. C. M. Starks, *J. Amer. Chem. Soc.*, 93, 195 (1971).
2. A. Brandstrom, "Preparative Ion Pair Extraction," an Introduction to Theory and Practice, Apotekarsocieteten, Hassle Lakemedel, First Ed., 1974; Second Ed., 1976.
3. A. Brandstrom, "Principles of Phase Transfer Catalysis by Quaternary Ammonium Salts," in "Advances in Physical Organic Chemistry," Vol. 15, V. Gold, Ed., Academic Press, London and New York, 1977, p. 267.
4. W. P. Weber and G. W. Gokel, "Phase Transfer Catalysis in Organic Synthesis," in *Reactivity and Structure*, Vol. 4, K. Hafner et al., Eds., Springer-Verlag, Berlin - Heidelberg, 1977.
5. C. M. Starks and C. Liotta, "Phase Transfer Catalysis. Principles and Techniques," Academic Press, London and New York, 1978.
6. E. V. Dehmlov and S. S. Dehmlov, "Phase Transfer Catalysis," First Ed., 1980; Second Revised Ed., Verlag Chemie, Weinheim, 1983.
7. L. J. Mathias and C. E. Carralier Jr., Eds., "Crown Ethers and Phase Transfer Catalysis in Polymer Science," Plenum Press, New York 1984.
8. V. Percec and B. C. Auman, *Makromol. Chem.*, 185, 617 (1984).
9. V. Percec and B. C. Auman, *Makromol. Chem.*, 185, 1867 (1984).
10. V. Percec, B. C. Auman and P. L. Rinaldi, *Polym. Bull.*, 10, 391 (1983).
11. V. Percec and H. Nava, *Makromol. Chem., Rapid Commun.*, 5, 319 (1984).
12. V. Percec, H. Nava and B. C. Auman, *Polym. J.*, 16, 681 (1984).
13. V. Percec, P. L. Rinaldi and B. C. Auman, *Polym. Bull.*, 10, 215 (1983).
14. V. Percec, P. L. Rinaldi and B. C. Auman, *Polym. Bull.*, 10, 397 (1983).
15. V. Percec and B. C. Auman, *Polym. Bull.*, 12, 253 (1984).
16. Percec, T. D. Shaffer and H. Nava, *J. Polym. Sci., Polym. Let. Ed.*, 22, 637 (1984).
17. T. D. Shaffer and V. Percec, *J. Polym Sci., Polym. Chem. Ed.*, 24, 451 (1986).
18. V. Percec, H. Nava and J. M. Rodriquez-Parada, *J. Polym. Sci., Polym. Lett. Ed.*, 22, 523 (1984).
19. V. Percec, H. Nava and J. M. Rodriquez-Parada, *Polym. Bull.*, 12, 261 (1984).
20. V. Percec, H. Nava and J. M. Rodriguez-Parada, in "Advances in Polymer Synthesis," J. E. McGrath and B. M. Culbertson, Eds., Plenum Press, New York, 1985, p. 235.
21. V. Percec and B. C. Auman, *Polym. Bull.*, 10, 385 (1983).

22. V. Percec, in "Cationic Polymerization and Related Processes," E. J. Goethals, Ed., Academic Press, London and New York, 1984, p. 347.
23. V. Percec and T. D. Shaffer, in "Advances in Polymer Synthesis," J. E. McGrath and B. M. Culbertson, Eds., Plenum Press, New York, 1985, p. 133.
24. T. D. Shaffer and V. Percec, Makromol. Chem. Rapid Commun., 6, 97 (1985).
25. T. D. Shaffer and V. Percec, J. Polym. Sci., Polym. Lett. Ed., 23, 185 (1985).
26. T. D. Shaffer, M. Jamaludin and V. Percec, J. Polym. Sci., Polym. Chem. Ed., 24, 15 (1986).
27. T. D. Shaffer and V. Percec, J. Polym. Sci., Polym. Lett. Ed., 24, 185 (1985).

RECEIVED August 20, 1986

Chapter 10

Mechanistic Aspects of Phase-Transfer Free Radical Polymerizations

Jerald K. Rasmussen, Steven M. Heilmann, Larry R. Krepski, and Howell K. Smith II

Corporate Research Laboratories, 3M, 3M Center, St. Paul, MN 55144

A critical survey of the literature on free radical polymerizations in the presence of phase transfer agents indicates that the majority of these reactions are initiated by transfer of an active species (monomer or initiator) from one phase to another, although the exact details of this phase transfer may be influenced by the nature of the phase transfer catalyst and reaction medium. Initial kinetic studies of the solution polymerization of methyl methacrylate utilizing solid potassium persulfate and Aliquat 336 yield the experimental rate law:

$$R_p = k[MMA]^{1.6}[Aliquat\ 336]^{0.2}[K_2S_2O_8]^{0.6}$$

This expression is consistent with theory in the limiting case where the equilibrium constant for anion exchange at the interface is large.

In 1981 we reported (2, 3) the first examples of free radical polymerizations under phase transfer conditions. Utilizing potassium persulfate and a phase transfer catalyst (e.g. a crown ether or quaternary ammonium salt), we found the solution polymerization of acrylic monomers to be much more facile than when common organic-soluble initiators were used. Somewhat earlier, Voronkov and coworkers had reported (4) that the 1:2 potassium persulfate/18-crown-6 complex could be used to polymerize styrene and methyl methacrylate in methanol. These relatively inefficient polymerizations were apparently conducted under homogeneous conditions, although exact details were somewhat unclear. We subsequently described (5) the

practical features of phase transfer free radical polymerization and some of the characteristics of the polymers formed. In the past few years, several additional accounts of phase transfer free radical polymerizations have appeared (6-12). These reactions have been described as being initiated by phase transfer of (Scheme 1):
a) ionic initiators from the solid phase into a liquid organic phase (2, 5, 6);
b) ionic initiators from a liquid aqueous phase into a liquid organic phase (3, 5, 11, 12);
c) neutral initiators from a liquid organic phase into a liquid aqueous phase (7-9); and
d) monomers from a liquid organic phase into a liquid aqueous phase (10).
This paper will focus on mechanistic aspects by reviewing the previous work in this field and by describing some recent kinetic studies conducted in our own laboratories.

Cyclodextrins as Phase Transfer Agents

Kunieda and coworkers, using cyclodextrins as phase transfer catalysts (7-9), have investigated the transfer of organic-soluble initiators into aqueous media for the purpose of polymerizing water-soluble monomers. Both heptakis(2,6-O-dimethyl)-β-cyclodextrin (DM-β-CD) and heptakis(2,3,6-O-trimethyl)-β-cyclodextrin (TM-β-CD) were found to accelerate the rate of polymerization of a variety of water-soluble monomers utilizing a variety of organic-soluble initiators in the two phase system water:(1:4 v/v) chloroform/ligroin (Scheme 2). Percent conversions ranging from 2.4 - 12.5 times those observed in the absence of cyclodextrin were recorded. In general, the results suggest that these lipophilic cyclodextrins may form inclusion complexes with the initiator in the organic phase, then transfer the initiator to the aqueous phase to allow polymerization to occur. The mechanistic details are not clear, however, since in only half of the cases studied was TM-β-CD, the more lipophilic catalyst, more efficient than DM-β-CD. Also, the two cyclodextrins responded quite differently to increasing polarity of the organic phase.

More recently, Kunieda has described (10) a new aspect of phase transfer free radical polymerization. In this study, methylated β-cyclodextrins were found to accelerate the polymerization of aryl group-containing water-insoluble monomers using water-soluble initiators in a two phase water/chloroform system. The results are consistent with transport of monomer from the organic phase to the aqueous phase, where initiation of the polymerization begins, followed by transport back to the organic phase for propagation. Again, however, more work will be necessary to unravel the exact mechanistic details.

Crown Ethers as Phase Transfer Agents

Takeishi, et. al, have described the redox polymerization of methyl methacrylate in the absence of solvent (6). With 18-crown-6 as the phase transfer catalyst and potassium persulfate/sodium bisulfite as the redox couple, polymerization was observed at temperatures <50°C whereas little or no polymerization occurred under these conditions in the absence of bisulfite. Above 55°C, however, polymerization occurred even in the absence of bisulfite. From the limited kinetic data reported (6), one can estimate (13) that the rate of polymerization (Rp) is approximately proportional to the square root of crown concentration (Equation 1):

$$Rp \; \alpha \; [18\text{-}Crown\text{-}6]^{0.6} \qquad (1)$$

In our initial studies of the polymerization of butyl acrylate by solid potassium persulfate in acetone solution (2), we attempted to relate the rate of polymerization to the ability of various crown ethers to complex the potassium cation. A reasonable correlation was discovered between log Rp and log K, where K represents the binding constant of the crown ether for K^+ in methanol solution (Figure 1). This finding provided some support for the idea that a typical phase transfer process was occurring in these reactions.

Quaternary Ammonium Salts as Phase Transfer Agents

Polymerization of butyl acrylate was also studied by us in ethyl acetate/water two phase systems (3) using potassium persulfate/quaternary ammonium salts as the initiator system. Under these conditions (a minimum amount of water was used to dissolve the persulfate), it was found that symmetrical quat salts were more efficient than surfactant type quat salts. Also, the more lipophilic quat salts were more efficient. These results prompted us to propose formation of an organic-soluble quaternary ammonium persulfate via typical phase transfer processes.

Until recently, the most detailed kinetic investigations of phase transfer free radical polymerizations were those of Jayakrishnan and Shah (11, 12). Both of these studies have been conducted in two phase aqueous/organic solvent mixtures with either potassium or ammonium persulfate as the initiator, and have corroborated our earlier conclusions (2, 3)

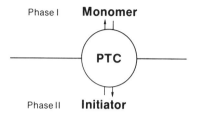

Scheme 1

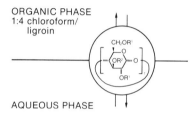

Scheme 2

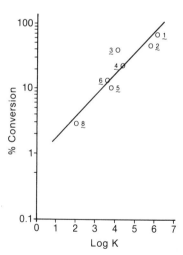

Figure 1. Dependence of butyl acrylate percent conversion to polymer on the stability constants for potassium ion complexation in methanol of the various crown ethers. Line calculated by regression analysis; o, experimental values; 1, 18-crown-6; 2, dicyclohexyl-18-crown-6; 3, 21-crown-7; 4, dibenzo-18-crown-6; 5, 15-crown-5; 6, cyclohexyl-15-crown-5; 8, 1,10-diaza-18-crown-6. Reproduced from Ref. 2. Copyright 1981, American Chemical Society.

concerning the increased efficiencies of phase transfer
catalyzed polymerizations in comparison to those
conducted using conventional initiators.
Polymerization of acrylonitrile (AN) in
toluene/water was studied using hexadecyltrimethylammonium bromide as the phase transfer catalyst (12).
In this case, a simple kinetic model could not be
derived to explain all the experimental observations.
Independently prepared hexadecyltrimethylammonium
persulfate was found to be soluble in toluene/AN
mixtures and to catalyze polymerization in this
homogeneous system at approximately the same rate as
that observed in the two phase system. This result
implies that anion exchange at the interface (see
below) must be essentially complete under these
conditions. Factors which complicated further analysis
of the mechanism included: a) precipitation of
poly(acrylonitrile) during the polymerization;
b) nonlinear percent conversion vs time,
especially decrease in polymerization rate with
time;
c) nonlinear rate vs concentration curves;
d) extremely slow or no polymerization at AN
concentrations <2.0 M; and
e) nonstirred reaction conditions.
Items b and d may be related since the decrease in rate
begins to be observed at about 40% conversion, i.e. at
the point where [AN] has fallen to about 2.25 M. It is
interesting to note that we (14) and others (15) have
also noted very slow rates of polymerization of ethyl
acrylate at 1 M concentrations in ethyl acetate using
either tetrabutylammonium bromide or 18-crown-6 as the
phase transfer catalyst. Further studies in this area
are needed.
The kinetics of methyl methacrylate (MMA)
polymerization in ethyl acetate/water two phase systems
was described as being more well-behaved (11). Using
hexadecylpyridinium chloride (HPC) as the phase
transfer catalyst, Rp was found to be approximately
first order in MMA concentration. In support of a
typical phase transfer mechanism, it was found that the
nature of the cation, NH_4^+ vs K^+, of the persulfate had
no effect on polymerization rate.
With respect to [HPC], it was observed that
although the rate of polymerization increased with
increasing quat salt concentration, there appeared to
be a tendency for the rate to level off as [HPC]
approached that of the persulfate. Log-log plots of
the data, however, are nearly linear and yield Rp α
$[HPC]^{0.4}$ at 60° (R=0.987) and Rp α $[HPC]^{0.65}$ at
54°C (R=0.978).
With respect to increasing persulfate concentration,
polymerization rate was observed to increase to a
maximum and then decrease. In this case, a log-log

plot showed no linear regions. Interestingly, Jayakrishnan and Shah observed that the presence of water was necessary for polymerization to occur in this system. Therefore it was postulated that this was not a true phase transfer catalyzed process, and that most of the quaternary ammonium persulfate was probably concentrated at the liquid-liquid interface. The possible involvement of micelles (16, 17), which have been shown to accelerate the rate of decomposition of persulfate in aqueous solution (16), was not investigated. Again the fact that these were unstirred reactions cloud the analysis of the results.

Very recently, Ghosh and Mandal have reported (20) a thorough kinetic investigation of the polymerization of styrene in the two phase system water/o-dichlorobenzene using potassium persulfate as initiator and tetrabutylammonium bromide as phase transfer agent. These studies were conducted at constant ionic strength and pH, and found that the rate of polymerization showed a square root dependence upon potassium persulfate concentration, tetrabutylammonium bromide concentration, and volume of the aqueous phase. The solubility of tetrabutylammonium persulfate in the organic phase was found to be quite low. Careful analysis of the results indicated that a significant part of the initiation was effected by decomposition of the persulfate in the aqueous phase followed by phase transfer of the sulfate radical anion into the organic/monomer phase.

New Results

Our kinetic studies have concentrated on the polymerization of MMA in organic solution using solid potassium persulfate and a phase transfer catalyst. All reactions were stirred at speeds in excess of 500 rpm to eliminate any effects due to stirring rate (17). Initially, polymerizations were conducted at 60°C in cyclohexanone solution. When 18-crown-6 was used as the catalyst, little difference was observed in the presence or absence of sodium bisulfite. Polymerization rates were in the range of 10-12% conversion per hour. Rp in the presence of methyltricaprylylammonium chloride (Aliquat 336), however, was approximately three times as fast. This result encouraged us to investigate the Aliquat system in greater detail.

Solvent polarity was found to effect the rate of polymerization (Table I). For the three solvents studied, an essentially linear relationship (R=0.970) was observed between Rp and dielectric constant.

Subsequent kinetic studies designed to elucidate the experimental rate law were conducted on o-xylene solution. Individual runs in general displayed outstanding linearity (Figure 2). The short induction period was determined to be due to the time required for the reaction mixture to reach thermal equilibrium;

Table I. Effect of Solvent Polarity on Polymerization Rate (a)

Solvent	Dielectric Constant (ε)	$10^4 R_p$ (mol l^{-1} sec^{-1})
Cyclohexanone	18.3	4.10
o-dichlorobenzene	9.93	3.53
o-xylene	2.57	2.30

(a) $60°C$; [MMA] = $3.75\underline{M}$; [$K_2S_2O_8$] = $9.35 \times 10^{-3}\underline{M}$; [Aliquat 336] = $1.79 \times 10^{-2}\underline{M}$

rates were extracted from the curves after equilibrium had been reached. The dependence of Rp on [MMA], [Aliquat 336], and [$K_2S_2O_8$] are listed in Tables II, III, and IV respectively. From log-log plots of the data, the experimental rate law for this range of conditions was found to be:

$$R_p = k[\text{MMA}]^{1.6}[\text{Aliquat } 336]^{0.2}[K_2S_2O_8]^{0.6} \quad (2)$$

The orders with respect to MMA and Aliquat 336 concentrations appear to be fairly accurate, showing correlation coefficients of >0.99. That for persulfate, R=0.923, is deemed somewhat less accurate.

Table II. Dependence (a) on [MMA] at $60°C$

[MMA] (mol l^{-1})	$10^4 R_p$ (mol l^{-1} sec^{-1})
1.88	0.787
2.34	1.03
2.82	1.37
3.32	2.09
3.75	2.30
4.69	3.23

(a) [Aliquat 336] = $1.79 \times 10^{-2}\underline{M}$; [$K_2S_2O_8$] = $9.35 \times 10^{-3}\underline{M}$ in o-xylene

Table III. Dependence (a) on [Aliquat 336] at $60°C$

10^3 [Aliquat 336] (mol l^{-1})	$10^4 R_p$ (mol l^{-1} sec^{-1})
4.48	1.67
8.98	1.98
17.9	2.30
35.9	2.51
53.9	2.78

(a) [MMA] = $3.75\underline{M}$; [$K_2S_2O_8$] = $9.35 \times 10^{-3}\underline{M}$ in o-xylene

Table IV. Dependence (a) on $[K_2S_2O_8]$ at 60°C

10^3 $[K_2S_2O_8]$ (mol l^{-1})	10^4 R_p (mol l^{-1} sec^{-1})
4.68	1.13
9.35	2.30
18.7	2.59

(a) [MMA] = 3.75$\underline{M}$; [Aliquat 336]=1.79x10^{-2}$\underline{M}$ in $\underline{o}$-xylene

Kinetic Expression

According to free radical addition polymerization theory, the rate of polymerization of monomer M is proportional to the square root of the initiator I concentration (Equation 3) when termination is bimolecular (18).

$$-d[M]/dt = R_p = k[I]^{0.5}[M] \quad (3)$$

In the phase transfer polymerization reactions, if one assumes a simple ion exchange at the interface (liquid-liquid or solid-liquid) (17) (Equation 4)

$$2(QX)_{org} + K_2S_2O_8 \overset{K_{eq}}{\rightleftharpoons} (Q_2S_2O_8)_{org} + 2KX \quad (4)$$

and that the quaternary ammonium persulfate is the initiator (little or no polymerization occurs in the absence of quat salt) it can be shown that:

$$[Q_2S_2O_8]_{org}^{0.5} = \frac{K_{eq}^{0.5}[Q]_{total}[K_2S_2O_8]^{0.5}[KX]}{[KX]^2 + 2K_{eq}[QX]_{org}[K_2S_2O_8]} \quad (5)$$

Hence,

$$R_p = k K_{eq}^{0.5} \frac{[Q]_{total}[K_2S_2O_8]^{0.5}[KX][M]}{[KX]^2 + 2K_{eq}[QX]_{org}[K_2S_2O_8]} \quad (6)$$

From Equations 4 and 6, two distinct situations can be envisioned:

Case 1 - K_{eq} is large

In this situation, $[KX]^2$ dominates the denominator and Equation 6 reduces to Equation 7:

$$R_p = k K_{eq}^{0.5} \frac{[Q]_{total}[K_2S_2O_8]^{0.5}[M]}{[KX]} \quad (7)$$

Case 2 - K_{eq} is small

Here, Equation 6 simplifies to Equation 8:

$$R_p = k \frac{[Q]_{total} [KX] [M]}{2 K_{eq}^{0.5} [QX]_{org} [K_2S_2O_8]^{0.5}} \quad (8)$$

Case 1 appears to accurately predict the observed dependence on persulfate concentration. Furthermore, as $[Q]_{total}$ approaches [KX], the polymerization rate tends to become independent of quat salt concentration, thus qualitatively explaining the relative insensitivity to [Aliquat 336]. The major problem lies in explaining the observed dependency on [MMA]. There are a number of circumstances in free radical polymerizations under which the order in monomer concentration becomes >1 (18). This may occur, for example, if the rate of initiation is dependent upon monomer concentration. A particular case of this type occurs when the initiator efficiency varies directly with [M], leading to Rp α $[M]^{1.5}$. Such a situation may exist under our polymerization conditions. In earlier studies on the decomposition of aqueous solutions of potassium persulfate in the presence of 18-crown-6 we showed (19) that the crown entered into redox reactions with persulfate (Scheme 3). Crematy (16) has postulated similar reactions with quat salts. Competition between MMA and the quat salt thus could influence the initiation rate. In addition, increases in solution polarity with increasing [MMA] are expected to exert some, although perhaps minor, effect on Rp. Further studies are obviously necessary to fully understand these polymerization systems.

Conclusion

In conclusion, several examples of free radical polymerizations under phase transfer conditions have been described in the literature since the initial reports in 1981. In all of these cases it is apparent that transfer of an active species from one phase to a second phase is intimately involved in the initiation step of the polymerization. However, it is also clear that these are complex reactions mechanistically, and one general kinetic scheme may not be sufficient to describe them all. The extent of phase transfer and the exact species transferred will depend to a large extent upon the nature of the two phases, upon the

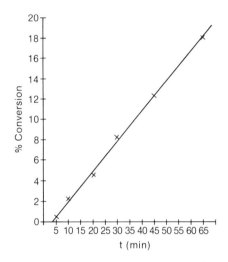

Figure 2. Percent conversion of MMA to polymer vs. time at 60°C; [MMA]=2.82M; [Aliquat336] =1.79x10^{-2}M; [K$_2$S$_2$O$_8$] = 9.35x10^{-3}M. Line calculated by regression analysis; X, experimental values (R=0.9995).

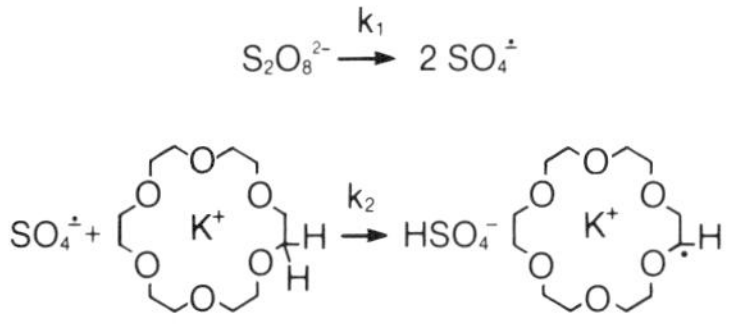

Scheme 3

polarity of the monomer being polymerized, and, quite importantly, upon the character of the phase transfer agent.

Experimental

All solvents used were reagent grade and were dried over 4A molecular sieves (21). Glassware was thoroughly washed, then rinsed three times with distilled-deionized water and oven-dried overnight prior to use. Potassium persulfate (Mallinckrodt Analytical Reagent) was recrystallized from distilled-deionized water and dried under vacuum prior to use. Methyl methacrylate (Aldrich Chemical Co.) was purified by passing it successively through silica gel and alumina, followed by distillation at reduced pressure. 18-Crown-6 (Aldrich Chemical Co.), sodium bisulfite (Fisher Scientific), and Aliquat 336 (Henkel Corp., equivalent weight 535 by chloride analysis) were used as supplied commercially.

Typical Procedure. Methyl methacrylate (18.76g, 0.187 Mol) and Aliquat 336 (0.480g, 8.97×10^{-4} Mol) were placed in a volumetric flask and diluted to 50 mL with o-xylene. A 250 mL 3-necked round bottomed flask equipped with a mechanical stirrer, condenser, and inert gas inlet was placed in a constant temperature bath at 60 ± 0.5°C. Potassium persulfate (0.1264g, 4.68×10^{-4} Mol) was placed in the flask and the contents purged with argon. Stirring (500 rpm) was begun and the monomer solution added. A 0.5 mL aliquot was withdrawn immediately as the zero point, and additional aliquots were taken at 5 to 15 minute intervals. The aliquots were quenched immediately by placing them in screw-capped vials maintained at 0°C (controls were run to verify that no polymerization occurred under these conditions over a period of several hours). Monomer concentration in the samples was determined by glpc (10% UCW 98 on Chromosorb W, column temperature 75°C) as the average of two trials, using the solvent as internal standard.

Literature Cited

1. This paper is part 7 in our series on free radical reactions under phase transfer conditions. For part 6, see reference (5).
2. Rasmussen, J. K., and Smith, H. K. II J. Am. Chem. Soc., 1981, 103, 730.
3. Rasmussen, J. K., and Smith, H. K. II Makromol. Chem., 1981, 182, 701.
4. Rakhmatulina, T. N., Baiborodina, E. N., Rzhepka, A. V., Lopyrev, V. A., and Voronkov, M. G. Vysokomol. Soedin., Ser. B, 1979, 21, 229; Chem. Abstr., 1979, 90, 187436v.

5. Rasmussen, J. K., and Smith, H. K. II In "Crown Ethers and Phase Transfer Catalysis in Polymer Science"; Mathias, L. J., and Carraher, C. E. Jr., Ed.; Plenum: New York, 1984; pp. 105-119.
6. Takeishi, M., Ohkawa, H., and Hayama, S. Makromol. Chem., Rapid Commun., 1981, 2, 457.
7. Kunieda, N., Taguchi, H., Shiode, S., and Kinoshita, M. Makromol. Chem., Rapid Commun., 1982, 3, 395.
8. Taguchi, H., Kunieda, N., and Kinoshita, M. Makromol. Chem., Rapid Commun., 1982, 3, 495.
9. Taguchi, H., Kunieda, N., and Kinoshita, M. Makromol. Chem., 1983, 184, 925.
10. Kunieda, N., Shiode, S., Ryoshi, H., Taguchi, H., and Kinoshita, M. Makromol. Chem., Rapid Commun., 1984, 5, 137.
11. Jayakrishnan, A., and Shah, D. O. J. Polym. Sci., Polym. Chem. Ed., 1983, 21, 3201.
12. Jayakrishnan, A., and Shah, D. O. J. Appl. Polym. Sci., 29, 2937.
13. Calculated from Fig. 1 of reference (6).
14. Smith, H. K. II, unpublished work from our laboratories.
15. Gozdz, A. S. Institute of Organic and Polymer Technology, Technical University of Wroclaw, Poland, private communication.
16. Crematy, E. P. Makromol. Chem., 1971, 143, 125.
17. Starks, C. S., and Liotta, C. "Phase Transfer Catalysis"; Academic Press, New York, 1978; Chapter 2.
18. Odian, G. "Principles of Polymerization"; Wiley-Interscience, New York, 1981; 2nd Edition, Chapter 3.
19. Rasmussen, J. K., Heilmann, S. M., Toren, P. E., Pocius, A. V., and Kotnour, T. A. J. Am. Chem. Soc., 1983, 105, 6845.
20. Ghosh, N.N., and Mandal, B.M. Macromolecules, 1986, 19, 19.
21. Burfield, D. R., Gan, G-H., and Smithers, R. H. J. Appl. Chem. Biotechnol., 1978, 28, 23.

RECEIVED July 26, 1986

Chapter 11

Aromatic Substitution in Condensation Polymerization Catalyzed by Solid-Liquid Phase Transfer

Raymond Kellman[1], Robert F. Williams[2], George Dimotsis[2], Diana J. Gerbi[2], and Janet C. Williams[2]

[1]Department of Chemistry, San Jose State University, San Jose, CA 95192
[2]Division of Earth and Physical Sciences, The University of Texas at San Antonio, San Antonio, TX 78285

> Solid-liquid phase transfer catalysis in nucleophilic aromatic substitution reactions is a novel approach to high molecular weight condensation polymers. This work has widened the scope of phase transfer catalysis and lead to a variety of new model compounds and high molecular weight polymers. Model aromatic oxide, sulfide, and imide nucleophiles have been used with a number of fluoronated aromatics as substrates in model reactions catalyzed by solid phase base and crown ethers. The analogous bis-nucleophiles react with these aromatics, and readily suffer disubstitution to afford high polymer. Reactivity in these heterogeneous systems is very sensitive to a variety of experimental conditions, esp. catalyst structure, solvent, and trace amounts of water in the liquid organic phase. These parameters greatly affect the concentration and structure of crown complexed cation-nucleophile ion pairs in solution. Evidence has been accumulated which suggests that an electron transfer rather than an anionic substitution mechanism operates especially with perfluoaryl substrates. Details of synthesis, optimization of PTC parameters and mechanisms are discussed.

The early, independent work of Starks, Markosa and Brandstrom from ca. 1965-1969, brought into focus with Starks' classical paper in 1971, showed PTC to be potent and versatile synthetic tool.[1-4] Since that time, the well-documented investigations of PTC have been massive and vigorous.[5-7] In polymer chemistry PTC was effectively exploited first in anionic addition polymerization and more recently has been extended to condensation polymerization.[5,8-9] However, until a very few years ago, the use of solid-liquid PTC systems in polycondensation has for the most part escaped this intensive scrutiny. Consequently, some time ago we began a rather broad study into the use of solid-liquid PTC to effect polycondensations.[10]

0097-6156/87/0326-0128$06.00/0
© 1987 American Chemical Society

We wish to report here some of the more recent results of our investigation.

Results and Discussion

We initiated our work by examining nucleophilic aromatic substitution, a somewhat difficult reaction to effect in other than activated aryl halides as substrates. It occurred to us that if polyhaloaromatics could be made to suffer disubstitution under mild solid-liquid PTC conditions, then they might be used as comonomers with a variety of bisnucliophiles to prepare halogenated polyarylethers, sulfides, sulfone-ethers as well as other interesting polymers which are at present synthesized only with some difficulty.

Our approach was to study structure reactivity relationships in a number of model reactions and, then, to proceed to the usually more difficult polymerizations using a variety of comonomer pairs. Secondly, we hoped to optimize the various, experimental solid-liquid PTC parameters such as nature and amount of catalyst, solvent, nature of the solid phase base, and the presence of trace water in the liquid organic phase. Finally, we wished to elucidate the mechanism of the PTC process and to probe the generality of solid-liquid PTC catalysis as a useful synthetic method for polycondensation.

Model studies using 4-isopropylphenol (1; X= -O-) and 4-t-butylthiophenol (1; X= -S-) with either hexafluorobenzene (2; HFB) or one of a number of bisaryl substrates were carried out in a solid-liquid PTC system which employed an appropriate organic solvent as liquid phase, a crown ether, usually 18-crown-6 as catalyst, and anhydrous K_2CO_3 as solid phase and base. The solid phase base reacts with the acidic phenols and thiophenols to form the reactive, crown-complexed potassium phenoxide and thiophenoxide nucleophiles, respectively. Reactions of 2 equivalents of the model phenol with HFB in refluxing acetone, DMAC, or acetonitrile afforded excellent yields (ca. 90%) of bis-1,4-(4-isopropylphenoxy)perfluorobenzene (3; X= -O-). Analysis by ^{19}F-NMR showed an isomer distribution of 99.8% para, and 0.1% meta substitution for the disubstituted products. Also, 0.1% of 1,2,4- and a trace of 1,3,5-trisubstituted (3; n=2) products were detected. In runs with more than 2 equivalents of the phenol 1,2,4,5-tetrasubstituted product (3; n=3) was found along with increased amounts of the trisubstituted products. The reaction of 2 equivalents of 4-t-butylthiophenol with HFB under similar conditions gave virtually quantitative yields of the corresponding bis-1,4-(4-t-butylthiophenoxy)perfluorobenzene (3; X= -S-). In this case, ^{19}F-NMR analysis showed exclusive para disubstitution. These model results demonstrated the feasibility of polycondensation under mild solid-liquid PTC conditions using HFB as difunctional comonomer and, further, that HFB would be incorporated into the polymer backbone via nucleophilic attack to give a largely parasubstituted, linear polymer. Formation of tri-and tetraryl- oxides in our models suggested that branching and/or crosslinking would occur along the backbone at the haloaromatic units in the fluoronated polyaryloxides.

Model studies were also carried out with several bisaryl substrates (4a-c); viz., perfluorobiphenyl (4a; PFB), perfluorophenyl sulfide (4b; PFPS) and perfluorobenzophenone (4c; PFBP). These reactions were run under the same PTC conditions as those with HFB. In every case, the bisaryls (4a-c) underwent 4,4'-disubstitution with 2 equivalents of nucleophile to give excellent, often quantitative yields of 5a-c as the only detectable ether or sulfide product. When excess nucleophile was employed, again only 5a-c were produced and no polysubstituted products were detected, which indicated that PTC polycondensation of bisaryls via aromatic substitution should be facile and lead to linear, branch free polymers. This proved to be the case.

Of particular interest was the reaction of two equivalents of potassium phthalimide with PFB using 18-crown-6 in refluxing acetonitrile. This reaction with either small molecules or the polymeric analogs represents a novel approach to arylimide synthesis via PTC. After 4 hr. under nonoptimized PTC reaction conditions, disubstitution afforded the bisimide 6 in ca. 50% yield. This shows that phthalimide anion, a considerably poorer nucleophile than either the phenoxide or thiophenoxide, is a strong enough nucleophile in the presence of 18-crown-6 to displace aryl fluoride with facility, and demonstrates that the synthesis of polyimides, an important class of thermally stable polymers, is feasible by this PTC polycondensation route.

Polymerizations were carried out using a number of bisphenols and bisthiophenols (7a-e) with HFB (see Scheme I) and with (4a-c) the bishaloaromatics (see Scheme II). Typically, solutions equimolar (ca. 2.5M) in comonomer were treated with excess anhydrous K_2CO_3 and 27.7 mole % of 18-crown-6 ether. Reactions were run in a number of different solvents and at temperatures from 55 to 80°. Our results showed that polymerizations were sensitive to PTC parameters such as solvent, catalyst, and trace amounts of water in the organic phase. Under optimized conditions the expected polymers were generally obtained in excellent yields and moderately high molecular weights as evidenced by n_{inh} measurements (see Table I).

TABLE I

Polymerization of Bisphenols and Bisthiophenol with Hexafluorobenzene(2)

Polymer	% Yield	n_{inh}[a]
8a	93	0.58 (CHCl$_3$)
8b	99	0.37 (THF)
8c	100	0.11 (THF)
8d	100	insol.
8e	84	insol.

a) .50 wt % solutions at 30°.

That branching/crosslinking occured with HFB was evidenced by rapid gelation in some runs, and by partial insolubility of some isolated polymers. Control experiments, runs without catalyst, were carried out to demonstrate the catalytic nature of the reaction for each

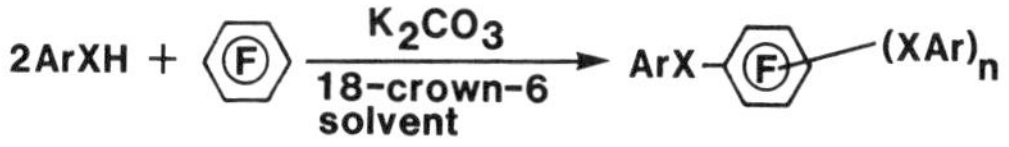

2ArXH + ⟨F⟩ $\xrightarrow[\text{solvent}]{\text{K}_2\text{CO}_3}$ ArX-⟨F⟩-(XAr)$_n$

1   2   3

X = -O-, -S-   n = 1-3

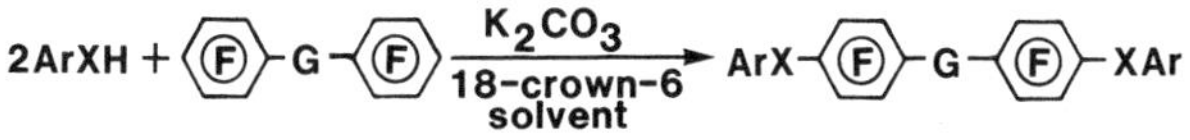

2ArXH + ⟨F⟩-G-⟨F⟩ $\xrightarrow[\text{solvent}]{\text{K}_2\text{CO}_3}$ ArX-⟨F⟩-G-⟨F⟩-XAr

4a-c   5 a-c

4a-c, G= −, -S-, -CO-

6

Scheme I

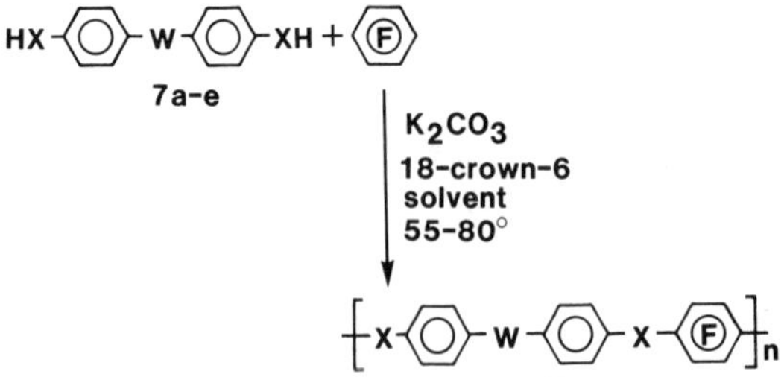

HX—⬡—W—⬡—XH + ⬡(F)
　7a-e

↓ K$_2$CO$_3$
　18-crown-6
　solvent
　55-80°

—[X—⬡—W—⬡—X—⬡(F)]$_n$
　　　　　8a-e

	X	W
8a	-O-	-C(CH$_3$)$_2$-
b	-O-	-S-
c	-O-	-SO$_2$-
d	-S-	-O-
e	-S-	-

Scheme II

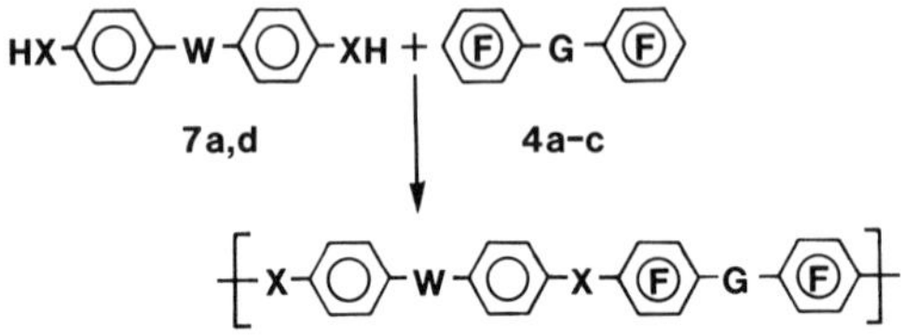

HX—⬡—W—⬡—XH + ⬡(F)—G—⬡(F)
　7a,d　　　　　　　4a-c

↓

—[X—⬡—W—⬡—X—⬡(F)—G—⬡(F)]—
　　　　　　9 a-f

	X	W	G
9a	-O-	-C(CH$_3$)$_2$-	-
b	-O-	-C(CH$_3$)$_2$-	-S-
c	-O-	-C(CH$_3$)$_2$-	-CO-
d	-S-	-O-	-
e	-S-	-O-	-S-
f	-S-	-O-	-CO-

different nucleophile and each solvent employed. All polymers were characterized by infrared spectroscopy and by elemental analyses. ^{1}H- and ^{19}F-NMR spectra were determined for all soluble polymers. Most of the HFB polymers (8a-e) showed good solubility in common solvents (e.g., THF, CHCl$_3$); whereas, most polymeric derivatives of the bishaloaromatics (9a-f), especially those with PFB as comonomer, were insoluble. Inherent viscosities ranging from 0.1 to 0.7 were obtained for soluble polymers. Coherent films could be either solution cast or melt pressed for a number of polymers.

A number of bisnucleophiles (7a-e) were polymerized with either PFB, BFPS and PFBP(4a-c). Reaction conditions were similar to those used for the HFB polymerizations. Most of the polymers obtained were insoluble materials and of low molecular weight. Exceptions to this trend were seen with bisphenol-A (7a) or 4,4'-dimercaptodiphenyl ether (7d) (see Table II). These nucleophiles reacted with bisaryls to give somewhat soluble polymers (9a-f) which, in turn, allowed conversion to high molecular weight (n_{inh} = 0.4-0.6).

TABLE II
Polymerization of Bisphenol-A and 4,4'-Dimercaptodiphenyl Sulfide with Fluoronated Bisaryls(4a-c)

Polymer	% Yield	n_{inh}[a]
9a	95	.62
9b	84	.22
9c	83	.21
9d	95	.46
9e	90	.29
9f	94	.31

a) .50 wt % solutions at 30°.

As anticipated, catalysis was strongly affected by polarity of the liquid phase (see Table III and IV). The catalytic effect was monitored by determining yields and viscosities of polymers formed under identical reaction conditions both with and without catalyst present. A small catalytic effect was observed in highly polar, aprotic media (e.g. DMAc) and a large catalytic effect in solvents of lower polarity (e.g. acetone). Acetonitrile provided an intermediate case. In highly polar media the ratio of concentrations of the simple nucleophilic ion pair to crown-complexed nucleophilic ion pairs is large and both the uncatalyzed and catalyzed processes make significant contributions to the overall rate of reaction. In solvents of low polarity the simple ion paired nucleophile is much less soluble, and this ratio of reactive nucleophilic species is small. Consequently, the rate of the uncatalyzed reaction is small and makes little contribution to the overall rate of aromatic substitution. However, the catalyzed rate is greatly enhanced in nonpolar media, due to the increased nucleophilicity of the crown-complexed ion pair. This results in a higher overall rate of substitution than is seen in the polar media[6]. With the exclusion of catalyst (i.e., control experiments) in solvents of lower polarity, the concentration of the simple nucleophilic ion pair, the

TABLE III
Effect of Solvent and Catalyst on Solid-liquid PTC
Polymerization of HFB[a]

Polymer	Solvent	% Yield	n_{inh}[c]
8a	DMAC	90	0.50
8a	DMAC[b]	88	0.42
8a	Acetone	93	0.58
8a	Acetone[b]	0	----
8a	DMSO	80	insol.
8a	DMSO[b]	50	insol.
8a	C_6H_5Cl	5	0.36
8a	C_6H_5Cl[b]	2	olig.
8a	Toluene	10	olig.
8a	Toluene[b]	3	olig.
8a	1,4-Dioxane	24	olig.
8a	1,4-Dioxane[b]	6	olig.
8a	THF	3	olig.
8b	DMAC	100	0.37
8b	DMAC[b]	0	----
8c	DMAC	100	0.11
8c	DMAC[b]	0	----
8d	DMAC	80	insol.
8d	DMAC[b]	84	insol.
8d	Acetone	84	insol.
8e	DMAC	93	insol.
8e	DMAC[b]	90	insol.
8e	Acetone	100	insol.
8e	Acetone[b]	0	----

a) Reactions were carried out at 80° in DMAC for 24 hr; 55° in acetone for 48 hr; 80° in chlorobenzene for 48 hr; 80° in DMSO for 24 hr; 80° in toluene for 48 hr; 80° in dioxane for 48 hr; and 65° in THF for 48 hr.
b) Control experiments carried out in the absence of catalyst.
c) .50 wt % solutions in $CHCl_3$ or THF at 30°.

only reactive nucleophilic species present, is small and responsible for the small or negligible overall rate of aromatic substitution.

TABLE IV
Effect of Solvent and Catalyst on Solid-liquid PTC
Polymerization of PFB[a]

Polymer	Solvent	% Yield	n_{inh}[c]
9a	DMAC	90	.50
9a	DMAC[b]	91	.42
9a	CH_3CN	95	.62
9a	CH_3CN[b]	66	.40
9d	DMAC	80	.30
9d	DMAC[b]	88	.28
9d	CH_3CN	95	.46
9d	CH_3CN[b]	92	.30

a) Reactions in DMAC at 80° and at reflux in CH_3CN
b) Control runs carried out in absence of catalyst
c) .50 wt % solution

High polymer was not formed in solvents such as chlorobenzene and toluene. Under PTC catalysis in such lipophilic, liquid media, high yields of oligomers, often dimers, trimers, and tetramers, were isolated and identified. In these model polymer insolubility precludes chain growth to high molecular weight species. Aryl polymers, especially with 1,4-phenylene or 4,4'-biphenylene units along the backbone, show low solubility and precipitate prematurely from solution. As expected, this problem was most acute for polymerization of the bisaryl substrates. Many of these materials were insoluble in all reaction solvents employed. Very significant, however, is the observation that PTC aromatic substitution does in fact, occur with facility even in these very weakly polar solvents.

The observed catalytic effect of the crown ether appears to be dependent on the nucleophile employed in both polymerization and corresponding model reactions. Not surprisingly, it appears that the stronger the nucleophile employed, the smaller the catalytic influence of the crown ether. For example, with potassium thiophenoxide yields of polymer or model products were almost quantitative with or without catalyst. By contrast, the reaction of PFB with potassium phthalimide, a considerably weaker nucleophile, affords 6 in 50% with catalyst and in 2-3% without catalyst under identical conditions. However, it may be that this qualitative difference in rates is, in fact, an artifact of different solubilities of the crown complexed nucleophiles in the organic liquid phase. A careful kinetic study of nucleophilicity in catalyzed versus non-catalyzed reactions study is presently underway.

The presence of trace amounts of water in the organic phase is known to affect profoundly the rate of solid-liquid PTC processes. [11-13] Only recently has this problem been addressed by work in our laboratory as well as by Liotta and Sasson. Early in our polymerization studies we found that PTC polymerization did not occur in "bone dry" or "wet" solvents. Consequently, we undertook a study of the effect of small amounts of water in the liquid, organic phase on the polymerization of bisphenol-A and HFB in several solvents.[14] Solvents were rigorously dried and assayed for water content by potentiometric Karl Fischer titration. A series of polymerizations catalyzed and uncatalyzed in each solvent were carried out in which the water content was increased incrementally. Polymer yields and inherent viscosities were determined as a function of water content in each solvent. The optimal water content expressed as the mole ratio of water to catalyst shows that the necessary water level varies substantially with solvent (see Table V and Figure 1-3).

TABLE V
Effect of Water on Polymerization of Bishpenol-A and HFB

Solvent	Optimal Mole Ratio H_2O/Catalyst[a]	% Yield	n_{inh}[b]
DMAC	1.7	90	0.50
Acetone	2.3	93	0.58
C_6H_5Cl	0.7	50	olig.

a) Water content determined by electronic Karl Fischer titrations.
b) .50 wt % solutions in $CHCl_3$ at 30°.
Source: Reproduced with permission from reference 14. Copyright 1985 John Wiley.

Although this effect was studied using the reaction of HFB and bishphenol-A as a model for fluoroarylether polymers, the influence of trace water also occurs in other monomer systems examined.[15] The enhancement of polymerization by trace water may be a result of solvation of the crown complexed ion pair which would increase the concentration of this species in a somewhat polar liquid phase. Alternatively, water may operate at the K_2CO_3 solid-liquid interface to facilitate ionization and crown-complexation of the solid base. Retardation of the polymeri- zation process at higher water concentrations may be a result of excessive solvation of the reactive nucleophile by protic water which would decrease its reactivity in an otherwise aprotic medium. Or as Liotta suggests,[13] the water may be localized on the surface of the solid salt particles and forms a new region in the solid- liquid PTC system. This region or "omega-phase" effectively extracts the 18-crown-6 from the organic liquid phase which would eliminate any catalyzed rate process that would otherwise occur in the liquid region.

An understanding of the mechanisms of solid liquid PTC aromatic substitution is not only fundamentally interesting but also serves as a useful guide toward extending this process as a general method of polycondensation. The mechanism of aromatic substitution is currently under re-evaluation.[16] Both experimental data and MINDO/3 calculations, which compare the preference of ipso substitution to hydrogen substitution on a benzene nucleus as a function of charge type, show ipso substitution to be preferred energetically by anionic intermediates, and to be energetically competitive with free radical intermediates.[17] Particularly interesting is the fact that fluorine suffers preferential substitution over other aryl

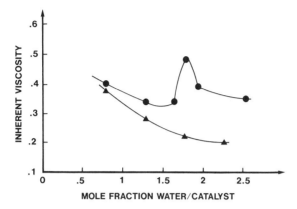

Figure 1. Effect of water on the preparation of polymer (8a) in DMAc in the presence (●) and absence (▲) of 18-crown-6 catalyst. (Reproduced with permission from reference 14. Copyright 1985 John Wiley.)

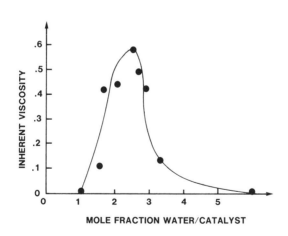

Figure 2. Effect of water on the preparation of polymer (8a) in acetone in the presence of 18-crown-6 catalyst. Uncatalyzed reactions produced no measurable polymeric products. (Reproduced with permission from reference 14. Copyright 1985 John Wiley.)

substituents in free radical aromatic substitution. Bunnett and Berasconi have shown that both S_NAr and $S_{RN}1$ mechanisms are general for nucleophilic aromatic substitution.[18-20]   We have previously reported results which implicate a free radical pathway in these solid-liquid PTC polycondensations as energetically competitive with anionic substitution in perfluoroaryl substrates (see Scheme III). [21-22]   That is, the ion pair complex of catalyst and nucleophile (11) may form a charge transfer complex(12a) with the electron deficient polyfluoroaryl nucleus. This complex, susceptible to electron transfer from nucleophile to substrate, leads to a transition state for aromatic substitution with a large degree of free radical character. Product formation(14) can result by loss of fluoride ion from the complex,(12b) and subsequent rapid collapse of the caged radical pair(13).

Experimental

Equipment. All melting points were determined with a Mel-Temp apparatus from Laboratory Devices and are uncorrected. Gas chromatography was carried out on either a Varian Aerograph 700 or Hewlett-Packard 5880 chromatograph. Beckman IR-9 and Perkin Elmer FT-1800 spectrophotometers were used for the determination of mid-IR spectra. All NMR spectra were obtained using an IBM NR-80 FT spectrometer. Elemental analyses were determined for all new monomers and polymers by Micro-Tech Laboratories, Inc. of Skokie, IL. Dilute solution viscosity measurements were done at 30° with the appropriate Ostwald-Fenske capillary viscometers. The water content of all organic solvents, used as the liquid phase in solid-liquid PTC runs was analyzed by potentiometric Karl Fischer titration using a Metrohm AG CH 9100 model automatic titrator.

Materials. All solvents used were reagent grade. Solvents were dried by known literature procedures, degassed, and stored under dry nitrogen. Bisphenol-A (Bis-A) and 4,4'-thiodiphenol (TDP) were recrystallized twice from toluene. 4,4'-Sulfonyl-diphenol (SDP) was recrystallized twice from a MeOH/water (65:35) mixture. 4-Isopropylphenol was recrystallized twice from hexane. 4,4'-Biphenyldithiol (BPDT) was recrystallized twice from an ethanol/water mixture. Hexafluorobenzene, perfluorobiphenyl (PFB), perfluorobenzophenone (PFBP), 4-t-butylthiophenol, 4,4'-dimercaptodiphenyl ether (DMDP), and 18-crown-6 ether (Parish) were used without further purification. Perfluorophenylsulfide (PFPS) was prepared by treating pentafluorobenzene with n-BuLi and SCl_2.[23]   Potassium carbonate (anhydrous) was oven dried (130°) overnight before use and transferred to reaction flask in a N_2 filled glove bag.

Model Studies.
Preparation of Bis-1, 4-(disubstituted)perfluorobenzene and Bis-4,4' (disubstituted)perfluorobiphenyl. To a solution of 4.85 mmole of HFB or PFB and 9.65 mmole of 4-isopropylphenol or 4-t-butylthiophenol in 20 ml of solvent was added 20.8 mmole of K_2CO_3 and 1.34 mmole of 18-crown-6 ether. The magnetically stirred heterogeneous mixture was heated to reflux or to 80° with DMAc and maintained under dry N_2 for 4-24 hr. For reactions with

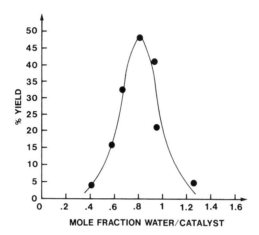

Figure 3. Effect of water on the reaction of hexafluorobenzene and bisphenol-A in chlorobenzene in the presence of 18-crown-6 catalyst. Products of the catalyzed reaction were oligomeric materials. No reaction is observed under the experimental conditions in the absence of the 18-crown-6 catalyst. (Reproduced with permission from reference 14. Copyright 1985 John Wiley.)

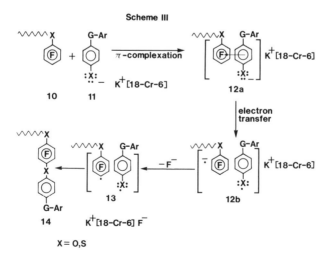

HFB, the reaction mixture was poured into 25 ml of ethyl ether and 50 ml of distilled water. The aqueous layer was discarded and the organic layer was washed twice with 25-ml portions of 10% aqueous NaOH and with 25 ml of a saturated NaCl solution. The ether solution was dried (MgSO$_4$) and concentrated (rotary evaporation) to give a crude product in ca. 90% yield. The crude products were recrystallized usually from ethanol or ethanol:H$_2$O to yield primarily the para-disubstitution products.

In the case of insoluble bis-4,4'-disubstituted perfluorobiphenyls the reaction mixtures were poured into ca. 100 ml of H$_2$O. The insoluble material was collected by filtration, and washed successively with H$_2$O, EtOH and CH$_2$Cl$_2$. The crude products were dried (vac. oven at 80°) and twice sublimed for purification. The model products were characterized by IR (neat liquid or KBr), mass spectrometry, and elemental analyses. ^{1}H and ^{19}F-NMR spectra we determined for all soluble products. Only minor modifications of this procedure were necessary with different haloaromatics or nucleophiles. For reactions of potassium phthalimide no K$_2$CO$_3$ was used or needed.

Polymerization Method. To a solution of 5.18 mmole of HFB or PFB and 5.18 mmole of the appropriate bisphenol or bisthiophenol in 20 ml of solvent was added 22.4 mmole anhydrous of K$_2$CO$_3$ and 1.43 mmole of 18-crown-6 ether. The magnetically stirred, heterogenous mixture was heated in an oil bath and maintained under N$_2$. Upon cooling to room temperature, the mixture was slowly poured into ca. 150 ml of methanol and was vigorously stirred. The filtered solids were washed three times in a blender with 300-ml portions of distilled water. The solids were air dried and subsequently placed in a vacuum oven (80°) for 24 hr. Where soluble, the polymers obtained were characterized by IR and PMR analysis. Elemental analyses for all polymers were satisfactory. Polymer solubility was determined in THF, DMF, dioxane, toluene, m-cresol, chloroform, and sulfuric acid. The percent insoluble polymer was determined gravimetrically. Inherent viscosities of soluble polymers were determined in ca. 0.5% wt. solutions in either chloroform or THF. For higher molecular weight polymers, films were cast from solution for soluble polymers and melt pressed ca. 300° for insoluble polymers.

Determination of the Effect of Water on Polymerization. To a solution of 4.76 mmole of HFB and 4.76 mmole of Bis-A in 20 ml of the appropriate solvent of known water content was added 20.6 mmole of K$_2$CO$_3$ and 1.32 mmole of 18-crown-6 ether. The crown ether, when necessary was dried by complexation with acetonitrile. Water content of solvent and catalyst was determined in all runs by Karl Fischer titration. Known microliter quantites of water in increments of 10µl were then added to the reaction mixture. The magnetically stirred, heterogeneous mixture was heated in an oil bath and maintained under N$_2$. Upon cooling to room temperature, the reaction mixture was slowly poured into ca. 300 ml of a nonsolvent vigorously stirred in a blender. The filtered solids were washed three times with 300-ml portions of distilled water. The product was air-dried and then placed under vacuum at 80° for 24 hr. Characterization was carried out as described above in "Polymerization Method".

Acknowledgments

RK wishes to acknowledge the San Jose State University Foundation and the Robert A. Welch Foundation of Houston, Texas (Grant AX-709) for their generous support of this work. In addition, RFW wishes to thank the Robert A. Welch Foundation of Houston, Texas (Grant AX-769) for their support.

Literature Cited

1. Makosa, M.; Serafinowa, B. Rocz. Chem. 1965, 339, 1223.
2. Brandstrom, A.; Junggren, U. Acta. Chem. Scand. 1959, 23, 2204.
3. Starks, C.; Napier, D. R. (to Continental Oil Co.), U.S. Patent 3,992,432 (1976); Netherlands Patent 6,804,687 (1968).
4. Starks, C. M. J. Am. Chem. Soc. 1971, 93, 195.
5. Weber, W. P.; Gokel, G. W. "Phase Transfer Catalysis in Organic Synthesis." Springer-Verlag, New York, 1978.
6. Starks, C. M.; Liotta, C. L. "Phase Transfer Catalysis: Principles and Practice", Academic Press, New York, 1978.
7. Dehmlow, E. V.; Dehmlow, S. S. "Phase Transfer Catalysis", 2nd Ed., Verlag Chemie, Weinheim, 1983.
8. Schnell, H. "Chemistry and Physics of Polycarbonate", Interscience Publishers, New York, 1964, pp. 37-41.
9. See "Symposium on Crown Ethers and Phase Transfer Catalysis in Polymer Chemistry", Polym. Prepr., Am. Chem Soc., Div. Polym. Chem. 1982, 23(1), 139-192 and references therein.
10. Kellman, R. et al. Polym. Prepr., Am. Chem. Soc., Div. Polym. Chem. 1980, 21(2), 164; ibid. 1981, 22(2), 383.
11. Kellman, R. et al. Polym. Prepr., Am. Chem. Soc., Div. Polym. Chem. 1981, 22(2), 385.
12. Zahalka, H. A.; Sasson, Y. J. Chem. Soc., Chem Commun. 1984, 1652.
13. Liotta, C. L. Prepr., Div. Pet. Chem., Am. Chem. Soc. 1985, 30(3), 367.
14. Kellman, R. et al. J. Poly. Sci., Poly. Lett. Ed. 1985, 23, 551.
15. Kellman, R. et al. Unpublished Results.
16. Traynham, J. G. Chem. Reviews 1979, 79(4), 323. and references therein.
17. Gandour R. D.; Traynham, J. G. "Abstracts of Papers", 173rd National Meeting of the American Chemical Society, New Orleans, La., March 1977; American Chemical Society; Washington, D. C.; Abstr. ORGN-137.
18. Orvik, J. A.; Bunnett, J. F. J. Am. Chem. Soc. 1970, 92, 2417.
19. Bunnett, J. F. Acc. Chem. Res. 1978, 11, 413.
20. Bernasconi, C. F. Acc. Chem. Res. 1978, 11, 147.
21. Kellman, R. et al. Polym. Prepr., Am. Chem, Soc., Div. Polym. Chem. 1981, 22(2), 387.

22. Kellman, R. et al., "Abstracts of Papers", 189th National Meeting of the American Chemical Society, Miami, Fl. April 1985; American Chemical Society; Washington, D. C.; Abstr. ORGN-100.
23. Tatlow, J. C. et al. J. Chem. Soc. 1964, 763.

RECEIVED July 26, 1986

Chapter 12

Triphase Catalysis in Organometallic Anion Chemistry

Robert A. Sawicki

Texaco Research Center, Beacon, NY 12508

The ability of both simple metal oxides and functionalized derivatives to assist in various chemical transformations is well documented. When these reactions are performed in the presence of a second insoluble reagent, a triphasic solid-liquid-solid reaction system can result. Immobilized polyethylene glycols have been shown to promote both displacement and carbonylation reactions. In those schemes where organometallic anions are produced, anionically activated alumina was found to be a superior reagent. Using the inorganic base/metal oxide combination, both dicobalt octacarbonyl and iron pentacarbonyl are converted to the corresponding metal anions. Reaction with carbon monoxide, alcohols and electrophiles such as benzyl halides and oxiranes produce the corresponding esters following a carboxyalkylation scheme. The advantages of this system include high selectivities, good yields, mild reaction conditions, and facile separations. Attempts have been made to identify the portion of the catalytic cycle which appears favorably influenced by the metal oxide surface.

The preparation of novel phase transfer catalysts and their application in solving synthetic problems are well documented([1](#)). Compounds such as quaternary ammonium and phosphonium salts, phosphoramides, crown ethers, cryptands, and open-chain polyethers promote a variety of anionic reactions. These include alkylations([2](#)), carbene reactions([3](#)), ylide reactions([4](#)), epoxidations([5](#)), polymerizations([6](#)), reductions([7](#)), oxidations([8](#)), eliminations([9](#)), and displacement reactions([10](#)) to name only a few. The unique activity of a particular catalyst rests in its ability to transport the ion across a phase boundary. This boundary is normally one which separates two immiscible liquids in a biphasic liquid-liquid reaction system.

0097-6156/87/0326-0143$06.00/0
© 1987 American Chemical Society

Recent efforts have concentrated on the immobilization of these materials onto both organic polymers(11) and metal oxides(12) to simplify, by filtration, the separation, recovery and recycle process. These supported catalysts now function in a triphasic environment in either a liquid-solid-liquid or solid-liquid-solid reaction mixture. In this case, the catalyst must transfer the anion from the surface of the crystal lattice to the liquid phase. Here adsorption phenomena often significantly affect the reaction rate(13).

Simple refractory oxides such as alumina have also been shown to promote chemical transformations(14). Displacement and oxidation reactions(15) as well as desulfurizations(16) and oxidative couplings(17) are enhanced through the use of impregnated supports. The main advantages observed in this area are again improved selectivities, reactivities, and simplified separations.

Phase transfer catalysis has more recently been applied to the important area of organometallic chemistry(18). Reported applications include both the preparation(19) and the use of transition metal catalysts in isomerizations(20), carbonylations(21) and reductions(22).

The first application of phase transfer catalysis in metal carbonyl chemistry was reported by Alper in 1977(23). It was found that metal carbonyl anions could be readily generated by this technique and used to prepare pi-allyl, cluster, and ortho-metalated complexes(24).

Our efforts in this area of catalysis began in 1980. Our initial emphasis was on the preparation of supported phase transfer catalysts. We later became interested in the chemistry of anionically activated alumina(25) and the reactivity of metal carbonyl anions prepared under these conditions. A brief description of our work in the preparation of these materials and their synthetic applications follows.

The potential of using phase transfer catalysis in solving large scale industrial chemical problems has resulted in an explosion of ideas in both the preparation and utilization of new materials. One such class of catalyst is the high molecular weight polyethylene glycols(26) which have been attached to organic polymers to allow for facile separations(27). This ability is of course most important in many industrial processes where the criteria set for catalyst selection includes recoverability as well as stability, availability and safety(28).

Our interest in polyethylene glycols centered on a simple scheme to immobilize these materials onto metal oxide surfaces. The surface of silica gel contains both silanol-OH groups and -O- strained siloxane groups(29). A simple synthetic pathway to produce covalently bonded glycols was proposed where reaction(30) would occur between the OH group of the glycol and the surface of a refractory oxide.

Results and Discussion

Immobilized Polyalkylene Glycols. Polyalkylene glycols and monomethyl ethers of various molecular weights were heated at toluene reflux with both silica gel and alumina. Carbon analyses of the products indicated the degree of functionalization to be between 0.1 and 0.4 mmoles of glycol per gram of support which is typical of the capacity of metal oxides. The powders are normally Soxhlet extracted with toluene and/or methanol to remove any adsorbed glycol prior to use. Several examples are given in Table I(31).

These materials were subsequently evaluated under phase transfer conditions to determine their catalytic ability(32). The results from the displacement reaction of potassium acetate with 1-bromobutane in various solvents are given in Table II. In toluene, the polyethylene glycol on silica showed the highest activity. The polypropylene glycol analog was however much less active. It has been reported in the literature(33) that polyethylene glycols develop a helical conformation that allows for their ability to complex metal cations. Polypropylene glycols, on the other hand, were found to have a non-planar zigzag chain because of the steric effect of the methyl group. Their low extracting power may be attributed to this conformational phenomena.

Table I. Polyalkylene Glycols Immobilized onto Metal Oxide Surfaces

Glycol(a)	Support	%C	mmoles glycol/gram
PEG-400	SiO2	6.9	0.33
PEG-400	Al2O3	4.1	0.20
PEGMME-350	SiO2	5.6	0.27
PEGMME-350	Al2O3	3.1	0.17
PPG-425	SiO2	6.0	0.30
PPG-425	Al2O3	3.8	0.19

(a) PEG-400 = polyethylene glycol, avg. mol. weight 400;
    PEGMME-350 = polyethylene glycol monomethyl ether, avg. mol. weight 350;
    PPG-425 = polypropylene glycol, avg. mol. weight 425.

Table II. Reaction of 1-Bromobutane with Potassium Acetate at Solvent Reflux

Catalyst	Conc(a)	Time(b)	Solvent	%Yield(c)
SiO2	--	3(6)	PhCH3	0(37)
PEG-400	10	3	PhCH3	32
PEG-400/SiO2	6.6	3(6)	PhCH3	58(70)
PEG-400/Al2O3	4.0	3	PhCH3	33
PPG-425/SiO2	6.0	3(6)	PhCH3	7(10)
PEG-400/SiO2	6.6	3	PhCH3/H2O	0
PEG-400/Al2O3	4.0	3	H2O	49

(a) Concentration as mmoles of glycol (calc).
(b) Time in hours either 3 or 6.
(c) Yields are based on gas chromatographic analysis using decane as the internal standard.

Polyalkylene glycols immobilized onto metal oxides appear most suited for reactions that do not require the transportation of an anion across a liquid-liquid phase boundary. The displacement reaction occurs readily in either an organic or aqueous reaction medium but none is observed when an aqueous-organic biphasic mixture is employed. Overall, the reactivity of these catalysts follows the solubility of the nucleophile in pure glycol. This was demonstrated experimentally by comparing the reactions of sodium iodide, potassium acetate, sodium acetate and potassium cyanide with bromobutane. The yields followed the salt solubility in pure glycol (NaI>KOAc>NaOAc>KCN).

Carbonylation of Benzyl Halides. Several organometallic reactions involving anionic species in an aqueous-organic two-phase reaction system have been effectively promoted by phase transfer catalysts(34). These include reactions of cobalt and iron complexes. A favorite model reaction is the carbonylation of benzyl halides using the cobalt tetracarbonyl anion catalyst. Numerous examples have appeared in the literature(35) on the preparation of phenylacetic acid using aqueous sodium hydroxide as the base and trialkylammonium salts (Equation 1). These reactions occur at low pressures of carbon monoxide and mild reaction temperatures. Early work on the carbonylation of alkyl halides required the use of sodium amalgam to generate the cobalt tetracarbonyl anion from the cobalt dimer(36).

$$C_6H_5CH_2X + CO + ROH + Base \xrightarrow{Co(CO)_4^-} C_6H_5CH_2COR + Base \cdot HX \quad (1)$$

As, for the most part, the corresponding ester derivatives are a more important synthetic target, recent literature has demonstrated methods to prepare the esters directly. Examples include the use of nickel carbonyl in a methanol/dimethylformamide solvent system(37); the direct conversion of benzyl alcohol to methylphenylacetate using cobalt carbonyl(38) and a reaction system which utilizes an ammonium salt bound to an organic polymer(39).

As the supported glycol catalysts worked better in promoting reactions in a single solvent system, we explored the direct carbonylation of benzyl halides using an alcohol solvent, base, and cobalt carbonyl. Our initial experiments concentrated on the reaction of benzyl bromide at room temperature and one atmosphere carbon monoxide. We chose sodium hydroxide as the base, methanol as the solvent, and looked at the product distribution. We were interested in the selectivity to ester and the reactivity of this system. The results are given in Table III.

Phenylacetic acid was found to be the main carbonylation product when sodium hydroxide was the base. The presence of a phase transfer catalyst, either the commercially available quaternary

Table III. Carbonylation of Benzyl Bromide in Methanol(a)

Base	Catalyst	PhCH2OCH3	PhCH2CO2CH3	PhCH2CO2H
NaOH	--	17	0	66
NaOH	BTEAC*	61	0	29
NaOH	PEGMME-350/Al2O3	75	0	25
NaOH/Al2O3	BTEAC*	42	3	21
NaOH/Al2O3	--	3	37	29

(a) Reaction conditions = room temperature, 24 hours, 1 atm CO, base (50 mmoles), benzyl bromide (25 mmoles), dicobalt octarbonyl (1.5 mmoles), and methanol (75 ml).
(b) Ether and ester yields were determined by VPC, acid yield was obtained following extraction and isolation.
* BTEAC = Benzyltriethylammonium chloride.

ammonium salt (BTEAC) or our glycol ether on alumina catalyst, favored the formation of the ether via a Williamson ether synthetic pathway(40). A report using iron pentacarbonyl in a similar scheme suggested that a two-phase liquid system was required to prevent the formation of the ether(41).

Anionically Activated Alumina. At this time we had also developed an interest in anionically activated alumina. These basic reagents were active in promoting alkylation(42), condensation(43) and hydrolysis(44) reactions. Thus, we impregnated alumina with sodium hydroxide and used this combination both with and without a phase transfer catalyst (benzyltriethyl ammonium chloride). When BTEAC was added, the conversion to ether was decreased and the formation of ester was noted. In the absence of a phase transfer catalyst, the ether became a minor product and methyl phenylacetate became the major product with coproduction of phenylacetic acid. This ester does not result from esterification of the acid as simple stirring of phenylacetic acid with NaOH/Al2O3 in methanol does not produce methyl phenylacetate.

To further explore this phenomenon, we prepared a variety of base on alumina reagents. Their activity in the carboxymethylation of benzyl bromide is presented in Table IV.

The reaction conditions were mild (room temperature, 1 atm CO) and a two-fold excess of base was used along with a catalytic amount of cobalt carbonyl. The product distribution was quantified by VPC. The mixtures contained starting material, ester product, and various amounts of methyl benzyl ether. No detectable amounts of benzyl alcohol, ketones, or hydrocarbons were seen. Potassium methoxide alone afforded mostly the ether. A mixture of potassium methoxide and alumina gave a slight improvement in ester yield but the predominant product was again the ether. In contrast, when potassium methoxide on alumina was used, the carboxyalkylated product, methyl phenylacetate, was prepared in 70 yield with little ether detected. Benzyl chloride reacted in a similar fashion under these mild reaction conditions. Other alkoxide and carbonate bases could be used as

Table IV. Carboxymethylation of Benzyl Bromide(a)

Base	Yield, %(b)
---	0
Al2O3	5
MeOK	21
MeOK+Al2O3	30
MeOK/Al2O3	70(53)
MeONa/Al2O3	74
EtONa/Al2O3	65
t-BuOK/Al2O3	45
Na2CO3/Al2O3	39(23)
NaHCO3/Al2O3	67

(a) Reaction conditions = room temperature, 1 atm CO, 24 hours, base (50 mmoles), benzyl bromide (25 mmoles), dicobalt octacarbonyl (1 mmole) and methanol (75 ml).
(b) Yields were determined by VPC using internal standard techniques, distilled isolated yields are given in parentheses.

well. The alkoxides could be prepared by reaction of sodium hydroxide with the alcohol and deposition of the mixture onto alumina.
Using this base (NaOMe/Al2O3) methyl phenylacetate was obtained in 74 percent yield. Also, other esters, such as isobutyl phenylacetate, could be prepared by reaction in the appropriate solvent (isobutanol, 67 percent yield) using the sodium alkoxide on alumina reagent(45).
The formation of byproduct methyl benzyl ether was the key reason for the low selectivity to ester in the absence of alumina. A more careful examination of the product distributions with time was made using the alkoxide, alkoxide on alumina and bicarbonate on alumina bases. The results from Table V indicate that the formation of ether was indeed the predominant pathway with alkoxide alone, while the presence of alumina retarded this conversion and promoted the carboxyalkylation pathway. The bicarbonate on alumina gave little ether product and excellent selectivity to the methyl phenylacetate.
The next series of experiments were run varying the base composition (Table VI). With only a slight excess of base, the yield of byproduct ether was much lower using potassium methoxide, benzyl bromide, and 1 atm CO. The alkoxide on alumina reagents again gave higher conversions and selectivities to ester.
A larger enhancement was observed using sodium bicarbonate. This largely insoluble base was very inactive in promoting the carboxyalkylation reaction even in the presence of a phase transfer catalyst. Simply mixing with alumina did not improve this activity. However, when deposited onto alumina, the resulting base was much more efficient and good yields and high selectivities to ester were noted. This ability does not appear to be restricted to alumna as it was demonstrated that other metal oxides such as silica and titania work equally well.

Table V. Carboxymethylation of Benzyl Bromide -
Selectivity to Ester(a)

Base	Time	Product Distillation(b)	
		Ester,%	Ether,%
MeOK	1 Hr	27	31
	4 Hrs	33	48
	24 Hrs	36	60
MeONa/Al2O3	1 Hr	56	12
	4 Hrs	74	13
	24 Hrs	74	13
NaHCO3/Al2O3	1 Hr	31	0
	4 Hrs	51	0
	24 Hrs	67	6

(a) Reaction conditions = room temperature, 1 atm CO, base (50 mmoles), benzyl bromide (25 mmoles), dicobalt octacarbonyl (1 mmole) and methanol (75 ml)
(b) Yields were determined by VPC using internal standard techniques, remaining material was starting halide.

Table VI. Carboxymethylation of Benzyl Bromide -
Ester vs Ether(a)

Base	Product Distribution(b)	
	Ester,%	Ether,%
MeOK	43	14
MeOK/Al2O3	55	3
MeONa/Al2O3	52	8
NaHCO3	15	3
NaHCO3+BTEAC	16	4
NaHCO3+Al2O3	25	4
NaHCO3/Al2O3	49	3
NaHCO3/PEGMME-350/Al2O3	50	2
NaHCO3/SiO2	54	2
NaHCO3TiO2	50	2
K2CO3*	23	33
K2CO3/Al2O3*	17	28

(a) Reaction conditions = room temperature, 1 atm CO, 4 hours, base (30 mmoles), methanol (75 ml), benzyl bromide (25 mmoles), dicobalt octacarbonyl (1 mmole).
(b) Yields were determined by VPC using internal standards.
* Iron pentacarbonyl (1 mmole) used as catalyst.

A recent report(46) on the use of iron carbonyl and potassium carbonate in a similar carboxyalkylation scheme to prepare methyl phenylacetate prompted us to examine the use of carbonate on alumina in a similar manner. It was suggested that if the amount of free base was less than the amount of iron carbonyl than ether formation would not occur being that iron carbonyl was a better electrophile than benzyl halide. Under our conditions, the metal carbonyl anion was formed but the yields of ester and selectivity was much poorer than when cobalt carbonyl was the catalyst.

Catalytic Cycle. Attempts to determine the reasons for the improved activity of the base on alumina reagents followed two paths. Being as solubility of base in methanol appeared to greatly effect the production of methyl benzyl ether, we compared the amount of extractable base with selectivity to ester. This concentration of base, presumably alkoxide, was determined by stirring the base in methanol, filtering, and titrating the filtrate with acid. This comparison is given as Table VII.

Table VII. Extractable/Titratable Base vs Carboxyalkylation Activity

Base	%Titrated(a)	Ester,%	Selectivity(b)
MeOK	67	43	57
MeOK/Al2O3	35	55	71
NaHCO3	0.3	15	75
NaHCO3/Al2O3	2.3	60	82
NaHCO3/SiO2	2.3	54	76

(a) The base (30 mmoles) and methanol (75 ml) were stirred at room temperature for fifteen minutes, filtered, and the filtrate titrated to the phenolphthalein endpoint with 1N HCl.
(b) Selectivity is defined as the percent of converted benzyl bromide which appears as methyl phenylacetate.

In the cases where substantial base can be termed extractable-titratable, i.e., methoxide, the deposition onto alumina results in a decrease in concentration of base and an increase in both yield and selectivity to methyl phenylacetate. In the bicarbonate case, the deposition results in an increase in available base and the yield of ester is increased dramatically.

It is postulated that the slow step in the carboxyalkylation of alkyl halides is the cleavage of the acyltetracarbonyl cobalt compound(47). These materials react slowly with alcohols at room temperature and more rapidly with alkoxide ion. It seems likely that the higher activity of the base on alumina reagents, can be related to their ability to supply a sufficient quantity of alkoxide to favor the carboxyalkylation step but not a quantity high enough to produce substantial amounts of ether. This can be further substantiated by observing the infrared spectrum of the reaction in progress.

Before addition of the benzyl halide, the only carbonyl adsorption peak is found at 1900 cm^{-1}, indicative of the cobalt tetracarbonyl anion. After addition, this band immediately disappears and peaks at 2000 cm^{-1} are observed. These most likely represent the corresponding acyl complex. Reaction with methoxide yields the product and regenerates the cobalt anion. In the absence of sufficient methoxide, the reaction requires attack by the much

poorer nucleophile(methanol). The complete catalytic cycle is shown (Equation 2).
The combination of base and alumina, in essence, represents a triphasic solid-liquid-solid catalytic system particularly when highly insoluble bases are used. While the results suggest that

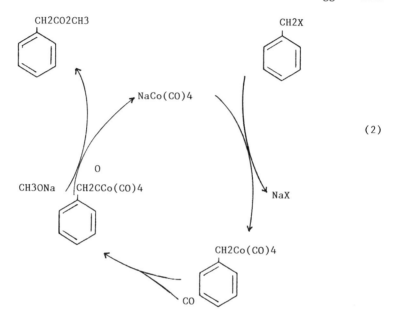

(2)

the key ingredient in increased activity is a solution phenomenon, the interaction of the substrate, catalyst or reaction intermediate with the metal oxide surface cannot be downplayed in importance.

Carboxyalkylation of Propylene Oxide. These reagents were also used in a similar carboxyalkylation scheme to prepare methyl 3-hydroxybutyrate by reaction with propylene oxide (Equation 3). This might represent a way to prepare substitute 1,3 diols(48) following reduction of the ester or reactive monomers by pyrolysis/dehydration.

Literature examples on the reaction of cobalt carbonyl(49), cobalt carbonyl anion(50), and iron carbonyl anion(51) with propylene oxide have been reported. However either the reaction conditions required are severe, the yields are low, or the reaction is not catalytic.

The results from our work on the reaction of propylene oxide with cobalt carbonyl and base in methanol are given in Table VIII. Several base/metal oxide combinations were evaluated under mild reaction conditions. The difference in activity between the bases was not as pronounced as that observed in the reaction with benzyl halides with the exception of potassium methoxide which, when used alone, gave exclusively the hydroxy ether resulting from methoxide addition to the epoxide ring. However, the activity of sodium bicarbonate did improve when the oxide was present. An explanation can be made by closer examination of the IR. In this system the predominant carbonyl adsorption band observed during reaction was that attributed to the cobalt anion. Here the slow step appears to be the addition of the cobalt tetracarbonyl anion to oxirane which

$$\underset{R'}{\triangle}\!\!-\!\!O + CO + ROH \xrightarrow{Co(CO)_4} \underset{R'}{\overset{HO\quad COR}{\underset{\|}{O}}} \quad \overset{2H_2/-ROH}{\swarrow} \quad \overset{-H_2O}{\searrow}\overset{\Delta}{\underset{R'}{\overset{O}{\underset{\|}{COR}}}} \quad (3)$$

is known to occur slowly at room temperature. There is not a build-up of acyl complex and thus less a dependence on available base. This system may be more representative of a typical reaction assisted by metal oxides where the increase in activity could be attributed to the adsorption of a substrate, catalyst or reaction intermediate.

Table VIII. Preparation of Methyl 3-Hydroxybutyrate(a)

Base	Ester,%(c)
NAHCO3	54
NaHCO3/Al2O3	74
NaHCO3/SiO2	70
NaHCO3/TiO2	62
MeOK/Al2O3	53
MeOK	0

(a) Reaction conditions = 6 hours, 75°C 50 psig, base (18 mmoles), propylene oxide (86 mmoles), methanol (75 ml) and dicobalt octacarbonyl (1.5 mmoles).
(b) Yields were determined by VPC using internal standard techniques.

Conclusions

The use of simple metal oxides and functionalized derivatives to solve the problems found in industrial chemical operations may be an important one. Any catalyst or reagent that is inexpensive, recyclable, separable, and allows reactions to occur more selectively and under milder reaction conditions would certainly be of interest to an already energy conscious industry. The wealth of information that is already available in this area through the efforts of both academic and industrial researchers speaks well of the potential for this technology.

Acknowledgments

I would like to thank Texaco Inc. for permission to publish this paper, J. Broas, Jr. for experimental assistance, and B. Townsend for preparation of this manuscript.

Literature Cited

(1) Dehmlow, E. V. and Dehmlow, S. S. "Phase Transfer Catalysis," Verlag Chemie, Weinheim, 1980; Starks, C. M. and Liotta, C. "Phase Transfer Catalysis, Principles and Techniques," Academic Press, New York, 1978; Weber, W. P. and Gokel, G. W. "Phase Transfer Catalysis in Organic Chemistry," Springer-Verlag, Berlin, 1977.
(2) Fedorynski, M.; Wojeiechowski, K.; Matacz, Z.; Makosza, M. J. Org. Chem. 1978, 43, 4682.
(3) Fujita, T.; Watanabe, S.; Suga, K.; Sugahara, K. Synthesis 1981, 1004.
(4) Dehmlow, E. V.; Barahona-Naranjo, S. J. Chem. Res.(S) 1981, 142.
(5) Guilmet, E; Meunier, B. Tetrahedron Lett. 1980, 4449.
(6) Imai, Y.; Kameyama, A.; Nguyen, T-Q.; Ueda, M. J. Polym. Sci. 1981, 19, 2997.
(7) Reger, D. L.; Habib, M. M.; Fauth, D. J. J. Org. Chem. 1980, 45, 3860.
(8) Alper, H.; Hachem, K.; Gambarotta, S. Can. J. Chem. 1980, 58, 1599.
(9) Nishikubo, T.; Iizawa, T.; Kobayashi, K.; Okawara, M. Tetrahedron Lett. 1981, 3873.
(10) Montanari, F.; Tundo, P. J. Org. Chem. 1981, 46, 2125.
(11) Molinari, H.; Montanari, F.; Quici, S.; Tundo, P. J. Am. Chem. Soc. 1979, 101, 3920.
(12) Tundo, P.; Venturello, P. Tetrahedron Lett. 1980, 2581.
(13) Montanari, F.; Landini, D.; Rolla, F. Top. Curr. Chem. 1982, 101, 147.
(14) Posner, G. H. Angew. Chem. Intl. Ed. Engl. 1978, 17, 487.
(15) Quici, S.; Regen, S. L. J. Org. Chem. 1979, 44, 3436.
(16) Alper, H.; Ripley, S.; Prince, T. L. J. Org. Chem. 1983, 48, 250.
(17) Jempty, T. C.; Miller, L. L.; Mazur, Y. J. Org. Chem. 1980, 45, 749.
(18) Alper, H. Fundam. Res. Homogeneous Catal. 1984, 4, 79.
(19) Gibson, D. H.; Hsu, W-L.; Ahmed, Fu. U. J. Organomet. Chem. 1981, 215, 379.
(20) Sasson, Y.; Zoran, A.; Blum, J. J. Mol. Catal. 1981, , 293.
(21) Brunet, J-J.; Sidot, C.; Caubere, P. Tetrahedron Lett. 1981, 1013.
(22) Alper, H.; Amaratunga, S. Tetrahedron Lett. 1980, 2603.
(23) des Abbayes, H.; Alper, H. J. Am. Chem. Soc. 1977, 99, 98.

(24) Alper, H.; des Abbayes, H. J. Organomet Chem. 1977, 134, C11.
(25) Keinan, E.; Mazur, Y. J. Am. Chem. Soc. 1977, 99, 3861.
(26) Gokel, G. W.; Goli, D. M.; Schultz, R. A. J. Org. Chem. 1983, 48, 2837.
(27) Kimura, Y.; Regen, S. L. J. Org. Chem. 1983, 48, 195.
(28) Reuben, B.; Sjoberg, K. CHEMTECH 1981, 315.
(29) Fritz, J. S.; King, J. N. Anal. Chem. 1976, 48, 570.
(30) Baverez, M.; Bastick, J. Bull. Soc. Chim. Fr. 1965, 3662.
(31) Sawicki, R. A. U.S. Patent 4,421, 675, 1983.
(32) Sawicki, R. A. Tetrahedron Lett. 1982, 2249.
(33) Yanagida, S.; Takahashi, K.; Okahara, M. Bull. Chem. Soc. Jpn. 1977, 50, 1386.
(34) Alper, H.; des Abbayes, H.; des Roches, D. J. Organometal. Chem. 1976, 121, C31.
(35) Cassar, L.; Foa, M. J. Organometal. Chem. 1977, 134, C15.
(36) Heck, R. F.; Breslow, D. S. J. Am. Chem. Soc. 1963, 85, 2779.
(37) Cassar, L.; Chiusoli, G. P.; Guerrieri, F. Synthesis 1973, 509.
(38) Imamoto, T.; Kusumoto, T.; Yokoyama, M. Bull. Chem. Soc. Jpn. 1982, 55, 643.
(39) Foa, M.; Francalanci, F.; Gardano, A.; Cainelli, G.; Umani-Ronchi, A. J. Organometal. Chem. 1983, 248, 225.
(40) Olson, W. T.; Hipsher, H. F.; Buess, C. M.; Goodman, I. A.; Hart, I; Lamneck, Jr., J. H.; Gibbons, L. C. J. Am. Chem. Soc. 1947, 69, 2451.
(41) Tanguy, G.; Weinberger, B.; des Abbayes, H. Tetrahedron Lett. 1983, 4005.
(42) Bram, G.; Fillebeen-Khan, T. J. Chem. Soc. Chem. Comm. 1979, 522.
(43) Muzart, J. Synthesis 1982, 60.
(44) Regen, S. L.; Mehrotra, A. K. Synth. Comm. 1981, , 413.
(45) Sawicki, R. A. J. Org. Chem. 1983, 48, 5382.
(46) Tustin, G. C.; Hembre, R. T. J. Org. Chem. 1984, 49, 1761.
(47) Kemmitt, R. D. W.; Russell, D. R. "Comprehensive Organometallic Chemistry, Vol. 5" Ed. Wilkinson, G. Pergamon Press, New York, 1982, pp 1-276.
(48) Murai, T.; Kato, S.; Murai, S.; Suzuki, S.; Sonoda, N. J. Am. Chem. Soc. 1984, 106, 6093.
(49) Eisenmann, J. L.; Yamartino, R. L.; Howard, Jr., J. F. J. Org. Chem. 1961, 26, 2102.
(50) Heck, R. F. J. Am. Chem. Soc. 1963, 85, 1460.
(51) Takegami, Y.; Watanabe, Y.; Masada, H.; Kanaya, I. Bull. Chem. Soc. Jpn. 1967, 40, 1456.

RECEIVED July 26, 1986

Chapter 13

The Scission of Polysulfide Cross-Links in Rubber Particles through Phase-Transfer Catalysis

Paul P. Nicholas

BFGoodrich Research and Development Center, Brecksville, OH 44141

Hydroxide ions transported by onium ions from water into swollen rubber particles rapidly break polysulfide crosslinks with little or no main chain scission. The reaction is not truly catalytic because the catalyst is consumed during the reaction. Model studies with N-methyl-N,N,N-tri-n-alkylammonium chlorides and di-2-cyclohexen-1-yl disulfide show that catalyst decomposition involves a highly selective demethylation of the quaternary ammonium ion by its 2-cyclohexenylthiolate counterion. This step is inhibited when certain alkylating agents are added. Rate studies of the Hofmann reaction show that it is too slow to be competative under these conditions. Several onium salts have been examined in this process. In general, those having several large alkyl substituents perform best, consistent with the known partitioning/reaction rate behavior of such catalysts in simple, low viscosity solvents. However, with N-methyl-N,N,N-tri-n-alkyl-ammonium chlorides, the rate decreases as the tri-n-alkyl substituents become very large (e.g., $C_{18}H_{37}$).

The declining commercial use of reclaim rubber is often attributed to its relatively poor mechanical properties compared to new rubber. This originates from the structural changes that occur during manufacturing (1). These processes cause extensive mechanical and chemical main chain scission to give highly branched chain segments that differ greatly from new rubber. The main goal of this study was to find low cost chemistry for converting scrap rubber into a material more nearly resembling the structure of new rubber. Specifically, we have examined phase transfer catalysis as a means of transporting hydroxide ions from water into rubber particles to cleave polysulfide crosslinks with little main chain scission.

A crosslinked rubber particle can be considered as a viscous hydrocarbon phase. In principle, phase transfer catalysis should apply to the chemistry of inorganic ions within such particles,

0097-6156/87/0326-0155$06.00/0
© 1987 American Chemical Society

though diffusion constants would necessarily be much lower than in simple, low viscosity solvents. However, swelling the particles with hydrocarbon solvents should increase diffusivities by several orders of magnitude (2). If hydroxide ions could be transported in this way at acceptable rates, then polysulfide crosslinks might be selectively broken through the dismutation reaction illustrated in Scheme I for disulfides (3). The fate of thiolate under phase transfer catalyzed conditions will be discussed later. A similar scheme should also apply to tri- and tetrasulfides, though interior sulfur atoms might ultimately be displaced as sulfide and/or bisulfide.

Results and Discussion

Scission of Polysulfide Crosslinks in Scrap Rubber Particles.
Throughout this study, we used a single lot of scrap rubber peelings having the average composition described in the Experimental Section. We began our studies with Aliquat 336 as the phase transfer catalyst because of its proven effectiveness in simple systems and its commercial availability. When devulcanization is performed in a refluxing benzene/aqueous NaOH mixture, the chemical crosslink density (M_{chem}) declines to a steady-state value in about 2 h. However, the extent of devulcanization is strongly dependent on the starting catalyst concentration (Figure 1). This suggests that the catalyst is being decomposed or otherwise rendered inactive during the reaction. A minimum value of 1.6×10^{-5} mol/g is achieved with 0.030 M Aliquat 336, about a 60% reduction in the total chemical crosslink density. As expected, this reaction is very slow in the absence of a swelling agent.

Scission of Polysulfide Crosslinks in a Model Crosslinked Polybutadiene. Figure 2 compares the molecular weight distribution of two soluble polybutadiene rubbers (BR). These were prepared by devulcanizing a model BR containing only polysulfide crosslinks, by both the phenyllithium (4) and the phase transfer catalyzed methods. The preparation of this vulcanized BR was first described by Gregg (4), who also showed that phenyllithium selectively breaks polysulfide crosslinks without breaking main chains. Since there is good agreement between these two distributions, we can conclude that both methods have similar selectivity. We find that the phase transfer catalyzed method also applies to model polyisoprene and poly(styrenebutadiene) rubbers prepared in the same way, though the extent of devulcanization is lower (~80%).

Origin of Catalyst Decomposition. The dependence of crosslink density reduction on catalyst concentration suggests that the catalyst decomposes during devulcanization. The Hofmann reaction (Equation 1) is a prime suspect under these conditions.

$$CH_3R_1R_2\underset{\underset{CH_2CH_2R_3}{|}}{N^+}OH^- \longrightarrow CH_3R_1R_2N + CH_2=CHR_3 + H_2O \qquad (1)$$

But does this reaction occur at a competitive rate? We find that the disappearance of the quaternary ammonium ion, Q^+, in benzene (^1H nmr)

SCHEME I

$$2RS-SR + 2OH^- \rightleftharpoons 2RSOH + 2SR^-$$

$$2RSOH \rightleftharpoons \overset{O}{\underset{\|}{R}S}-SR + H_2O$$

$$\underline{\overset{O}{\underset{\|}{R}S}-SR + 2OH^- \rightleftharpoons RSO_2^- + SR^- + H_2O}$$

OVERALL $2RS-SR + 4OH^- \rightleftharpoons RSO_2^- + 3SR^- + 2H_2O$

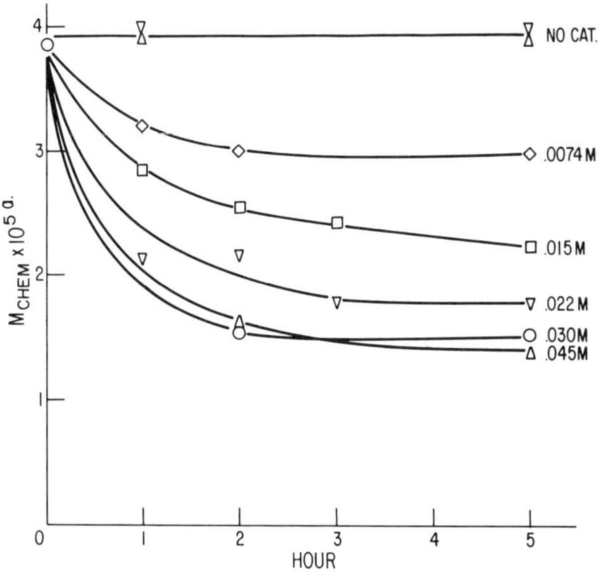

$M_{CHEM} \times 10^5$ a.

HOUR

a. Chemical Crosslink Density, mol/g Rubber

Figure 1. Scission of polysulfide crosslinks in rubber peelings with refluxing mixture of 2.1 N NaOH and Aliquat 336 in benzene.

under devulcanization conditions but in the absence of rubber obeys a first-order rate law (Figure 3). The apparent rate constants and half-lives are compared in Table I. The apparent constant k_a would contain not only the absolute rate constant(s) but also equilibrium constants and sodium hydroxide concentration terms. It is

Table I. Apparent Rate Constants for Hofmann Elimination with Aliquat 336

N, NaOH	k_a, h^{-1}	k_a, rel	$t_{1/2}$, h	T, °C
12.15	0.40	3.6	1.7	74.1
9.66	0.11	1.0	6.1	72.7
2.10	~0	~0	>>6.1	

possible to rationalize a first-order rate law, especially for the case where equilibration of ion pairs is rapidly established between the two phases. However, it is difficult to account for the high apparent order with respect to the aqueous sodium hydroxide concentration. As expected, the disappearance of Q^+ is accompanied by the formation of the corresponding amine in 89% yield. These results clearly show that Hofmann elimination is much slower than devulcanization, which is essentially complete in about 2 h with 2.1 N NaOH under these conditions.

We next considered the possibility that the quaternary ammonium ion might be dealkylated by the thiolate ion derived from the crosslink. Deno has reported examples of dealkylation in certain quaternary ammonium thiophenoxides (5). To test this possibility, we applied the devulcanization reaction to di-2-cyclohexen-1-yl disulfide (1), a simple model of a disulfide crosslink. This model has a disulfide attached to an allylic, cis-double bond, in general agreement with the crosslink structure of accelerated sulfur crosslinked rubbers (6). We performed the reaction under devulcanization conditions but at a disulfide concentration greater than the chemical crosslink density determined in the scrap rubber particles. The disappearance of disulfide was followed by HPLC. Figure 4 shows that the disulfide 1 is rapidly and essentially quantitatively decomposed to 3-(methylthio)cyclohexene (4). The broken curves define the theoretical methyl sulfide concentration based on the stoichiometry of the disulfide dismutation reaction. We believe, therefore, that demethylation is both the origin of catalyst decomposition and the ultimate fate of thiolate ions in this reaction (Scheme II). Since the observed stoichiometry of this reaction fits that of the disulfide dismutation reaction, we assume that the remaining cyclohexenethio fragment has been converted to the sulfinate 3.

Though N-methyl quaternary ammonium salts are not true catalysts in this reaction, only small amounts are needed because of the low concentration of crosslinks (typically <10^{-4} mol/g). Interestingly, we observe exclusive methyl transfer from the quaternary ammonium ion and no detectable transfer of R, where R = n-C_6H_{13} or mixed C_8-C_{10} alkyl. Considering our detection limits, we estimate a lower limit

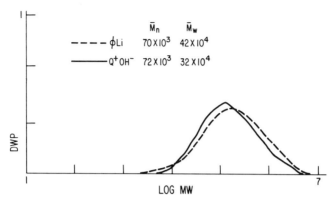

Figure 2. Molecular weight distribution for soluble rubber produced from the devulcanization of a model BR vulcanizate.

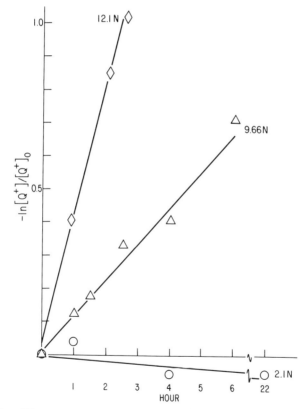

Figure 3. Disappearance of Aliquat 336 by Hofmann elimination in refluxing benzene/aq NaOH.

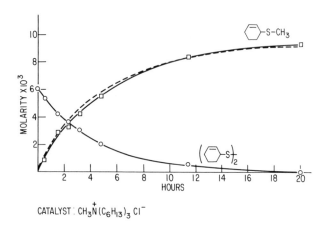

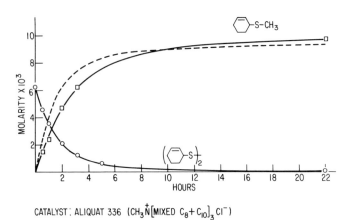

Figure 4. Disulfide bond scission in di-2-cyclohexen-1-yl disulfide.

of 260 for k_{CH_3}/k_R, indicating a highly selective alkylation reaction. Deno has also reported preferred demethylation with certain quaternary ammonium thiophenoxides but with much lower selectivity (5). The good agreement between the theoretical and observed methyl sulfide concentrations with time suggests that demethylation is fast compared to sulfur-sulfur bond breaking and the concentration of the thiolate 2 must be low.

We find that demethylation can be inhibited with 2% v/v of alkylating agent, such as benzyl chloride, added to the benzene phase (Figure 5). In this case, we observe only 3-(benzylthio)cyclohexene (6), though it is not produced quantitatively. Apparently, added excess alkylating agent can be more competitive than the quaternary ammonium ion in alkylating the thiolate 2, thereby freeing the quaternary ammonium ion to continue the catalysis (Equation 2).

$$2 + PhCH_2Cl \longrightarrow \langle\!\!\!\bigcirc\!\!\!\rangle\text{-S-CH}_2\text{-}\langle\!\!\!\bigcirc\!\!\!\rangle + CH_3R_3N^+Cl^- \quad (2)$$

<u>Devulcanization in the Presence of Benzyl Chloride and Methyl Chloride.</u> The above results suggest that catalyst efficiency might be improved when devulcanization is carried out with added alkylating agent. We find that this is, indeed, the case. Added benzyl chloride or methyl chloride further decreases the crosslink density for a given concentration of catalyst (Table II). However, 1- and 2-chlorobutanes appear to be ineffective, possibly because of dehydrochlorination.

Table II. Effect of Alkylating Agent on Devulcanization[a]

Alkylating Agent	Solvent	Volume, %	Reaction Time, h	$M_{chem} \times 10^5$
PhCH$_2$Cl	Benzene	20	5	1.0
PhCH$_2$Cl	Benzene	2	5	1.1
None	Benzene	0	5	1.6
PhCH$_2$Cl	Toluene	2	2	1.1
None	Toluene	0	2	1.4
PhCH$_2$Cl	p-Xylene	2	2	1.0
None	p-Xylene	0	2	1.5
CH$_3$Cl	p-Xylene	b	2	1.0
CH$_3$Cl	Benzene	b	5	1.5
CH$_3$Cl	Toluene	b	2	1.2
1-Chlorobutane	Benzene	2	5	1.4
2-Chlorobutane	Benzene	2	5	1.6

a. 0.030M Aliquat 336, b. Sat'd. with CH$_3$Cl vapor.

SCHEME II

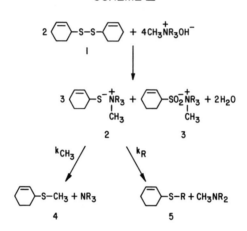

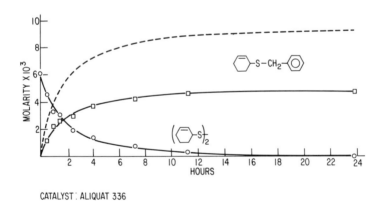

CATALYST: ALIQUAT 336

Figure 5. Disulfide bond scission in di-2-cyclohexen-1-yl disulfide.

Thus, about 74% of the chemical crosslinks are broken in this way. Most of those that remain likely originate from monosulfide and carbon-carbon bonds.

Other Onium Salt Catalysts. We have examined several quaternary ammonium and phosphonium salts in this process. And while they vary in their effectiveness, most seem to have at least some activity. In the case of symmetrical onium salts, substituents having four or more carbon atoms are preferred (Table III). This likely originates from a more favorable partitioning into the organic phase as the size of the substitutents increase. Herriott has reported that there is a

Table III. Other Onium Salt Catalysts[a]

Y	R	X	Solvent	$M_{chem} \times 10^5$
$X^{-+}YR_4$	N	C_2H_5	Br	Benzene	3.3
$X^{-+}YR_4$	N	C_3H_7	Br	Benzene	2.8
$X^{-+}YR_4$	N	$n-C_4H_9$	Br	Benzene	1.7
$X^{-+}YR_4$	N	C_2H_5	Br	Toluene	2.3
$X^{-+}YR_4$	N	C_3H_7	Br	Toluene	2.1
$X^{-+}YR_4$	N	$n-C_4H_9$	Br	Toluene	1.9
$X^{-+}YR_4$	P	C_2H_5	Cl	p-Xylene	2.3
$X^{-+}YR_4$	P	$n-C_4H_7$	Cl	p-Xylene	1.9
$X^{-+}YCH_3R_3$	N	$n-C_4H_9$	Cl	Benzene	1.7
$X^{-+}YCH_3R_3$	N	$n-C_5H_{11}$	Cl	Benzene	1.4
$X^{-+}YCH_3R_3$	N	$n-C_6H_{13}$	Cl	Benzene	1.4
$X^{-+}YCH_3R_3$	N	$n-C_8H_{19}$	Cl	Benzene	1.6
$X^{-+}YCH_3R_3$	N	$n-C_{12}H_{25}$	Cl	Benzene	1.8
$X^{-+}YCH_3R_3$	N	$n-C_{18}H_{37}$	Cl	Benzene	2.5

a. Two hour reaction times, 72°C, 2.1 N NaOH.

good correlation between phase transfer catalyzed reaction rates and phase partitioning of the onium ion in simple, low viscosity solvent (7). However, we also observe an unusual rate retarding effect as the alkyl substituents in methyl-tri-n-alkylammonium chlorides become very large, e.g., R = $n-C_{18}H_{37}$. In this case, we may be observing a slow, diffusion controlled reaction that is peculiar to the polymeric organic phase and might not occur in low viscosity solvents.

Experimental Section

Devulcanization of Rubber Peelings. Rubber peelings (10 g)(8) were charged into a 3-neck, 250 mL flask fitted with a water-cooled condenser and paddle stirrer. A solution of catalyst in 50 mL of benzene was then added. The mixture was stirred, heated to gentle reflux, and a solution of 2.0 g (.050 mol) of NaOH in 24 mL of water was added. The mixture was then vigorously stirred for 2 h. It was cooled, poured into 1.5 L of water, stirred for 15 min, and filtered. Washing in this way was repeated until the pH of the wash solution was 7-8. The rubber was then dried in a vacuum oven overnight at 55°C.

Crosslink Density Determinations. Approximately 1.4 g of rubber peelings was weighed into 120 mL screw cap bottles. About 80 mL of benzene was then added and the mixture allowed to stand to remove extractables. A single, 24 h extraction is sufficient. Following extraction, most of the solution was decanted off and the swollen rubber transferred to tared, seamless 85 mL aluminum cans which come with tight fitting lids (Ellisco, Inc., Pennsauken, NJ). A copper screen prepared from such a lid was placed over the can and the residual solvent shaken out. The inverted can was then struck several times against a paper towel to remove as much solvent as possible. The samples were then transferred to a vacuum oven and dried at 55° to constant weight (~2 h). The cans were placed in a glass vapor swelling chamber containing benzene, a sample tray, lead ballast, and an impeller mounted in the cover which is stirred with an external magnet. A few boiling chips were added to the benzene, and a vacuum was applied until the benzene boiled vigorously. The chamber was then filled with nitrogen to restore atmospheric pressure. This was then repeated. An open stopcock was directed to a mercury well through a short length of Neoprene rubber tubing. The chamber was then placed in a constant temperature bath at 25.5°C and the impeller turned at 1000 rpm. The following day, the stopcock to the mercury well was closed, and vapor stirring continued for a total of 48 h. The chamber was then opened, the lids quickly placed on the cans, and the samples weighed. Once covered, we found that weight loss due to benzene desorption was slow. No fewer than duplicate measurements were made.

The absolute chemical crosslink density values for the rubber peelings are only approximate because the treatment was applied to a rubber blend. However, the data are internally consistent and useful for comparing changes that occur with devulcanization. Chemical crosslink densities (M_{chem}, mol/g rubber hydrocarbon) were calculated from swelling data using the Flory-Rehner equation (9) and the Moore-Watson calibration curve (10) that correlates physical crosslink density with chemical crosslink density. We used a weighted average of 0.39 for the interaction parameter μ based on the rubber composition of the peelings. Before applying the Moore-Watson calibration, the chain end correction was made according to Mullens (11) using an estimated value of 177,000 for $\overline{M}n$ and a correction for carbon black as described by Porter (12). Rubber, carbon black, and ash content were determined by pyrolytic methods.

BR Containing Disulfide Crosslinks. This crosslinked BR was prepared using the procedure described by Gregg (4). Cured sheets were ground with dry ice and sieved to >18 mesh. Sieving is very difficult because of excessive static charge. Devulcanization was carried out under nitrogen at reflux for 4.5 h using 2.5 g of rubber, 75 mL benzene, 0.39 g Aliquat 336, and 6 mL of 2.1 N NaOH. At this point, the mixture comprised a somewhat viscous cement. There was little visual evidence of gel and no particles were observed. Methyl iodide (5 mL) was added, the mixture cooled to room temperature, and then filtered through a no. 100 copper screen (0.15 mm openings). This retained 0.29 g (12%) of dried rubber gel. The filtrate was again filtered (with difficulty and frequent filter replacement) through Whatman no. 41 paper. The total, retained gel (dried) was an additional 0.29 g. A small amount of an oxidation inhibitor was added and the sample analyzed by gpc.

A sample of the sieved particles was also devulcanized by the phenyllithium method (4). An oven-dried 200 mL beverage bottle was charged with 1 g of 18-60 mesh particles and 40 mL of dry benzene. The bottle was capped with a 1-hole cap (extracted neoprene liner) and 10 mL of 2 M phenyllithium added. The mixture was shaken for 6.5 h at room temperature, then 1.5 mL of methanol added. Finally, 4 mL of methyl iodide was injected. At thi point the mixture comprised a rubber solution and a suspension of salts. An aliquot of the solution was taken for gpc analysis.

Hofmann Elimination with Aliquat 336. A 2 L, 3 neck flask, creased on four sides, was fitted with a paddle stirrer, water cooled condenser, and a 3-way stopcock for sampling with a syringe. The flask was charged with 500 mL of benzene and 7.50 g (.0149 mol) of Aliquat 336 (CH_3NR_3Cl, R = C_8 and C_{10} chains, 442 avg. mol. wt., General Mills Chemicals, Inc., 88% or gtr. active). The stirred solution was heated to reflux and 240 mL of aqueous NaOH added. The mixture was vigorously stirred at reflux for about 10 min before timing began. Stirring was periodically interrupted long enough to remove a sample from the benzene phase. With runs requiring few samples (high rates), 12 mL samples were taken while 7 mL samples were taken for lower rates. Solvent was removed with a rotary evaporator and either 1.00 or 0.50 mL of a 0.00180 M solution of hexamethyldisiloxane in benzene was added as an internal standard. The disappearance of the alpha-protons of the ammonium ion $[CH_3N(CH_2-)_{3}]^+$ was followed by ^1H nmr. The combined peak areas were normalized to the area of the standard. They gave a broad resonance with a maximum intensity at δ 2.02. The amine accounted for 89% of the converted ammonium ion at the conclusion of the 9.66 N NaOH run.

Di-2-cyclohexene-1-yl disulfide (1). This synthesis is based on the general procedure described by Mattes and Chapman (13). A solution of 74.9 g (0.465 mol) of 3-bromocyclohexene, 38.9 g (0.512 mol) of thiourea and 232 mL of ethanol were refluxed for 3 h at 82°C. A solution of 41.0 g of NaOH in 256 mL of water was then added and the mixture refluxed for an additional 3 h. In a separate flask, a mixture of 86.4 g (0.536 mol) of 3-bromocyclohexene, 536 mL of methanol, 153 g (0.615 mol) of $Na_2S_2O_5 \cdot 5H_2O$ was refluxed for 2 h, after which time the solution became clear. This solution was then added dropwise to the first solution with stirring. After about 30 min at 30°C, 400 mL of a 50% (v/v) ether-pentane solution was added and the mixture stirred overnight at room temperature. The organic layer was removed and the water phase extracted twice with 75 mL of ether-pentane solution, which was then combined with the original organic layer. The combined organic layers were then washed twice with 100 mL of saturated aqueous sodium chloride and finally dried for 3 h over Na_2SO_4. The solvent was then removed by distillation, with the last trace removed under aspirator vacuum. The crude product (81.3 g) was purified using a Waters Preparative Liquid Chromatograph, Prepak 500 silica cartridge, 1% methylene chloride in hexane solvent, with 20 mL injections. The instrument was operated in the recycle mode with a center cut of the main peak recycled three times. The solutions were concentrated on a rotary evaporator and dried in a vacuum oven at 40°C overnight to give 47.7 g of >90% pure material. Further purification was achieved by low temperature crystallization from 10% v/v ethanol in hexane under N_2. NMR ($CDCl_3$) δ 1.99 (m, 12) δ 3.50 (broad S, 2) 3.3 (weak shoulder, δ 5.81 (m, 4).

Anal. Calcd. for $C_{12}H_{18}S_2$: C, 63.66; H, 8.01; S, 28.32. Found: C, 62.15; H, 7.71; S, 29.29.

A small second component was observed in the liquid chromatograph (Waters Bondapak C_{18} column, 60/4 acetonitrile/water solvent). Since a d,ℓ and meso form are possible, both may be present.

2-Cyclohexene-1-thiol. This compound was prepared according to the procedure described by Wolf and Paugh (14): bp 113°C (152 mm), 48.3%; lit. bp 108-112°C (182 mm). NMR ($CDCl_3$) δ 5.70 (m, 2), δ 3.47 (broad m, 1), δ 2.0 (v. broad m) + δ 1.65 (d, J = 4 Hz) = 7.

3-(Alkylthio)cyclohexenes. These reference compounds were prepared by the same general procedure. A 125 mL Erlenmeyer flask containing a magnetic stirring bar was charged with 50 mL of methanol and 5.00 g (0.0438 mol) of 2-cyclohexen-1-thiol. The solution was cooled in an ice bath and 1.0 g (0.044 mol) of sodium added with stirring. After the sodium had reacted, 0.438 mol of alkylating agent was added (methyl iodide, n-hexyl bromide, or benzyl chloride) and the solution stirred at room temperature for 16 h. Solvent was removed with a rotary evaporator, and the crude product shaken with 50 mL of water. The organic phase was separated, dried over $CaSO_4$, then vacuum distilled.

    1. 3-(Methylthio)cyclohexene. Bp 94°C (49 mm). NMR ($CDCl_3$) δ 1.75 (m) + δ 2.07 (s)(9), δ 3.23 (broad s, 1) δ 5.72 (m, 2).

    2. 3-(n-Hexylthio)cyclohexene. Bp 95°C (0.75 mm). NMR ($CDCl_3$) δ 0.87 (d, J = 5 Hz) + δ 1.25-1.97 (m, 17), δ 2.52 (t, 2, J = 7 Hz), δ 3.32 (broad s, 1), δ 5.72 (m, 2).

Anal. Calcd. for $C_{12}H_{23}S$: C, 72.66; H, 11.18; S, 16.16. Found: C, 71.87; H, 11.57; S, 15.99.

    3. 3-(Benzylthio)cyclohexene. Bp 103°C (0.04 mm). NMR ($CDCl_3$) δ 1.45 (m, 6), δ 3.23 (broad s, 1), δ 3.70 (s, 2), δ 5.68 (m, 2), δ 7.24 (m, 5).

Anal. Calcd. for $C_{13}H_{17}S$: C, 76.42; H, 7.89; S, 15.69. Found: C, 76.42; H, 7.90; S, 15.97.

Phase Transfer Catalyzed Decomposition of Di-2-cyclohexen-1-yl Disulfide (1). A 3-neck, 1 L flask, creased on four sides, was fitted with a thermometer, water-cooled condenser directed to a gas bubbler, a stopcock containing a rubber septum, and a mercury seal stirrer. The flask was purged with N_2 and charged with a solution of 0.525 g (0.00232 mol) of disulfide 1, 0.00695 mol of N-methyl-N,N,N-trialkyl-ammonium chloride and 350 mL of benzene. The solution was added through the condenser while a gentle flow of N_2 was continually passed through the system. The system was maintained under a positive pressure of N_2 and heated to reflux with stirring. A degassed, 2.1 N solution of aq. NaOH was then added (175 mL). Refluxing (68°C) resumed in about 4 min and this was considered zero time. Sampling was performed by stopping the stirrer, allowing the phases to separate, and removing about 1-1.5 mL from the benzene layer with a syringe. This operation required about 20-25 sec. A 2 µl sample was then injected into a Varian 850 HP liquid chromatograph using a Vari-Chrom Variable UV detector at 225 nm, a Waters 3.9 mm x 30 cm Bondapak C_{18} column, and a solvent comprising 40% (v/v) water in acetonitrile. The flow rate was 45 mL/hr with an increase to 90 mL/hr following the elution of 3-(methylthio)cyclohexene. The instrument had been calibrated with authentic samples using peak heights relative to benzene, the internal standard. Experiments with added benzyl chloride were carried out with 2% (v/v) of the benzene substituted with benzyl chloride.

Acknowledgments

Several people contributed to this program, both experimentally and through stimulating discussions. My thanks to J. R. Beatty, M. L. Dannis, E. C. Gregg, D. J. Harmon, E. R. Hooser, W. J. Kroenke, A. J. Magistro, J. B. Pausch, D. B. Ross, A. T. Schooley, P. J. Spear, J. Westfahl, C. E. Wilkes, and a special thanks to Eugene Licursi who performed most of the experimental work described in this report.

Literature Cited

1. Reviews of reclaim processes: (a) D. S. LeBeau, Rubber Chem. Technol. 1967, 40, 217; (b) U.S. Environmental Protection Agency, "Rubber Reuse and Solid Waste Management", U.S. Government Printing Office, Washington, D.C., 1971, p. 48; (c) A. Nowry, Ed., "Reclaim Rubber, Its Development, Applications, and Future", Maclaren and Sons, London, 1962.

2. G. J. van Amerogen, Rubber Chem. Technol. 1964, 37, 1065.

3. M. Calvin, "Mercaptans and Disulfides", Atomic Energy Commission, UCRL-2438, January 14, 1954, p. 27, and references cited therein.

4. E. C. Gregg, Jr. and S. E. Katrenick, Rubber Chem. Technol. 1970, 43, 549.

5. M. Shamma, N. C. Deno, and R. F. Remar, Tetrahedron Letters 1966, 1375.

6. L. Bateman, Ed., "The Chemistry and Physics of Rubber-Like Substances", John Wiley and Sons, New York, 1963, p. 451.

7. A. W. Herriott and D. Picker, J. Am. Chem. Soc. 1975, 97, 2345.

8. Average composition of benzene extracted particles: 58% rubber, 38% carbon black, 3.6% ash, 1.62% sulfur, NR/SBR/BR = 59/22.5/18.5; particle size 86% 30-60 mesh, 14% >60 mesh.

9. P. J. Flory and J. Rhener, Jr., J. Chem. Phys. 1943, 11, 512.

10. C. G. Moore and W. F. Watson, J. Polym. Sci. 1956, 19, 237.

11. L. Mullins, J. Appl. Polymer Sci. 1959, 2, 1.

12. M. Porter, Rubber Chem. Technol. 1967, 40, 866.

13. K. C. Mattes, O. L. Chapman, and J. A. Klun, J. Org. Chem. 1977, 42, 1815.

14. J. Wolf and T. Paugh, Rubber Chem. Technol. 1968, 41, 1329.

RECEIVED August 12, 1986

Chapter 14

Multisite Phase-Transfer Catalysts

John P. Idoux[1] and John T. Gupton[2]

[1]Lamar University, Beaumont, TX 77710
[2]University of Central Florida, Orlando, FL 32816

> The syntheses of a variety of "multi-site" phase-transfer catalysts (PTCs) and the determination of their catalytic activity toward some simple Sn2 reactions and some weak nucleophile-weak electrophile SnAr reactions are described. In general, at the same molar ratio, the "multi-site" PTCs are as or more effective than similar "single-site" PTCs. Thus, the "multi-site" PTCs offer an economy of scale compared to "single-site" PTCs.

Phase transfer catalysis (1,2) has become in recent years a widely used, well-established synthetic technique applied with advantage to a multitude of organic transformations. In addition to a steadily increasing number of reports in the primary literature, there are several reviews (3-6), comprehensive monographs (7-10) and an ACS Audio Course (11) which describe the phase transfer process and which provide extensive compilations of phase transfer agents and reaction types. While the list of applications and in many cases the synthetic results are impressive, phase transfer catalysts (PTCs) suffer some of the same disadvantages as more conventional hetero- and homogeneous catalysts -- i.e., separation and recovery. In the former case, because of the nature of most PTCs used in organic transformations, contamination of the desired product can be a problem (e.g., the relatively inexpensive quaternary ammonium and phosphonium salts often form stable emulsions.) In the latter case, because of the cost of some of the most efficient PTCs (e.g., cryptands and crown ethers), economy of scale can be a major consideration. In principle, these difficulties could be overcome (9,12) by attaching the PTC to an insoluble polymeric support; and, in fact, there has been a considerable amount of interest in this area including reports on polymer-bound quaternary salt PTCs (13,14), polymer-bound cryptand and crown ether PTCs (15,16) and at least one review on the subject (17).

In all cases, previously reported polymer-bound PTCs are ones which contain one PTC site/functionalized arm of polymer. Because of the ability to vary the nature of the substrate introduced onto the polymer arm, the preparation of a polymer-bound PTC with more than one PTC site/functionalized arm of polymer becomes possible. In theory, then, the number of grams of polymer backbone material needed to carry a particular level of required PTC active-site equivalency would be less for a "multi-site" substance compared to previously reported polymer-bound "single-site" PTCs. Similar considerations apply as well to nonpolymeric "multi-site" PTCs derived from simple, polyhalo substrates. Therefore, in general, "multi-site" PTCs offer the potential (i) of providing greater PTC activity on a PTC site/g of PTC needed for catalytic activity basis and (ii) of effecting a particular synthetic transformation under milder and/or more efficient conditions. Thus, economy of scale and efficiency are important considerations for both polymeric and nonpolymeric "multi-site" PTCs.

Multi-Site PTC Activity In Some Sn2 Reactions

In consideration of the above points, we have previously reported (18) the syntheses of two insoluble, polymer-supported, multi-site PTCs (II and III, Scheme 1) and a limited study of their effectiveness in two simple Sn2 reactions. The impetus for this investigation was provided in part by a report by Reeves' (13), who in contrast to reports (19) at that time by other workers, had demonstrated that it was not necessary to separate the PTC site of a polymer-supported PTC from the polymer backbone by long (i.e., 8-39 atoms) spacer chains in order to achieve activity. For example, Reeves' reported (13) that the polystyrene-backbone material I

$$[P] - (CH_2)_3 - PBu_3^+Br^-$$

I

was as or more active toward a variety of Sn2 reactions than considerably longer chain, polymer-supported PTCs (18). On the other hand, both Reeves (13) and Tomoi (20) have also reported that polystyrene-supported PTCs with $(CH_2)_x$ spacer chains of x = 3 or x = 4-7, respectively, are more active in Sn2 reactions than related PTCs where x = 1-2. Our preliminary report (18) on multi-site PTCs confirmed these observations. We report here the synthesis of an additional polymer-supported, multi-site PTC, the syntheses of several soluble, multi-site PTCs and the results of catalytic utility studies of these PTCs toward a number of simple Sn2 reactions.

The polymer-supported, multi-site PTCs shown in Scheme 1 are three-carbon analogs of I and were

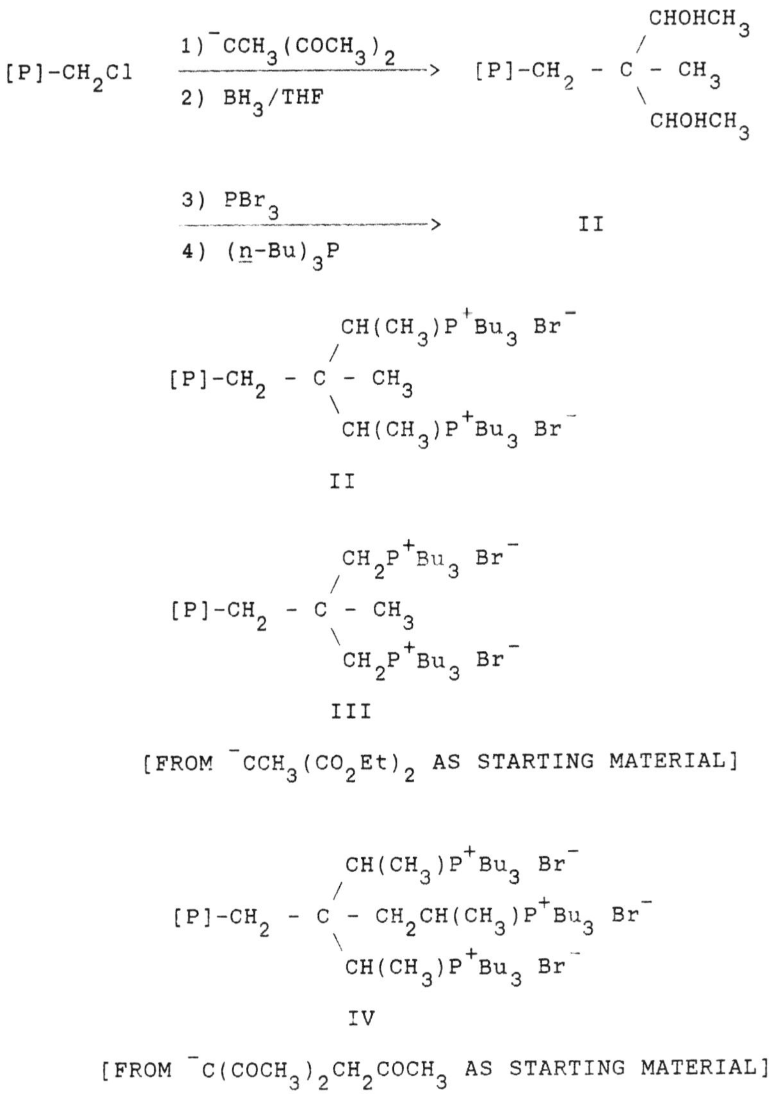

$$[P]-CH_2Cl \xrightarrow[\text{2) } BH_3/THF]{\text{1) } {}^-CCH_3(COCH_3)_2} [P]-CH_2-\underset{\underset{CHOHCH_3}{\diagdown}}{\overset{\overset{CHOHCH_3}{\diagup}}{C}}-CH_3$$

$$\xrightarrow[\text{4) } (\underline{n}\text{-Bu})_3P]{\text{3) } PBr_3} \quad II$$

$$[P]-CH_2-\underset{\underset{CH(CH_3)P^+Bu_3 \; Br^-}{\diagdown}}{\overset{\overset{CH(CH_3)P^+Bu_3 \; Br^-}{\diagup}}{C}}-CH_3$$

II

$$[P]-CH_2-\underset{\underset{CH_2P^+Bu_3 \; Br^-}{\diagdown}}{\overset{\overset{CH_2P^+Bu_3 \; Br^-}{\diagup}}{C}}-CH_3$$

III

[FROM ${}^-CCH_3(CO_2Et)_2$ AS STARTING MATERIAL]

$$[P]-CH_2-\underset{\underset{CH(CH_3)P^+Bu_3 \; Br^-}{\diagdown}}{\overset{\overset{CH(CH_3)P^+Bu_3 \; Br^-}{\diagup}}{C}}-CH_2CH(CH_3)P^+Bu_3 \; Br^-$$

IV

[FROM ${}^-C(COCH_3)_2CH_2COCH_3$ AS STARTING MATERIAL]

SCHEME 1

synthesized from commercially available chloromethyl polystyrene, [P]-CH$_2$Cl (obtained from Bio-Rad as a 2% divinylbenzene cross-linked material containing 4.04 meq Cl/g polymer). The first step in each sequence is the preparation of a polymer-supported carbonyl material obtained by reaction of the chloromethylated polymer with the requisite carbonyl alpha-carbanion. The polymer-supported carbonyl material is then reduced to the corresponding polymer-supported alcohol via reaction with BH$_3$/THF complex, followed by conversion to the bromide and finally reaction with tri-n-butyl phosphine to yield the desired multi-site phosphonium PTC. Combustion analyses indicated that polymers II, III and IV contained in excess of 90% of the theoretical number of meq of phosphorus/g of polymer. Compound V, the soluble, monomeric

$$PhCH_2CCH_3(CH_3CHPBu_3^+Br^-)_2$$

V

equivalent of II, was similarly synthesized previously (18) via the sequence shown in Scheme 1 from benzyl chloride as starting material (Anal. Calcd for C$_{37}$H$_{72}$Br$_2$P$_2$: P, 8.40%. Found: P, 8.31%). In an effort to evaluate the multi-site methodology in a variety of other structural types, several additional soluble, multi-site PTCs (VI-XIII) were synthesized from the corresponding polyhalo compounds via reaction with tri-n-butyl phosphine. The various aliphatic di-site (VI-XI),

$$(CH_2)_x (PBu_3^+Br^-)_2$$

VI-XI
x = 3, 4, 5, 6, 9, 12

the aromatic di-site (XII)

$$Ph[C(CH_3)_2PBu_3^+Cl^-]_2$$

XII
1,4-Disubstitution

and the aromatic tetra-site (XIII)

$$Ph[C(CH_3)(PBu_3^+Br^-)(CH_2PBu_3^+Br^-)]_2$$

XIII
1,4 - Disubstitution

materials all gave acceptable carbon, halogen, hydrogen and phosphorus analyses.

In order to determine the efficacy of the catalytic abilities of the various multi-site PTCs and to compare those abilities to those of previously reported PTCs, we have used the same test reaction and conditions reported by Reeves (13) for polymer-supported PTC I. Thus, the results shown in Table I for reaction of 1-bromopentane with aqueous solutions of various nucleophiles allow comparisons to be made among the catalytic abilities of the multi-site PTCs and those of other materials. As expected, the reactions proceed only minimally under "no catalyst" or under DMSO co-solvent conditions. As indicated, excellent yields of product are obtained via catalysis by PTCs II-IV through three reaction cycles and, in addition, are superior to those obtained via catalysis by PTC I. In general, catalysis via the non-polymeric, multi-site PTCs V-XIII gave good yields of recovered products and represent reactions under classical PTC conditions (i.e., one run and catalyst not recovered). In addition, the activities of the various multi-site PTCs compare favorably with those of several different soluble PTCs [aza-macrobicyclic polyethers (21), hexadecyltributylphosphonium bromide (21-22), benzyl tributylammonium halides (23) and methyl tricaprylammonium chloride (24)] as well as with those of other polystyrene-supported PTCs (13,19(a),(b),(d),25).

Thus, the various multi-site PTCs appear to offer a number of particularly attractive considerations. They are relatively easy to prepare (PTCs VI-XIII from readily available polyhalides) and PTCs II-IV, like other polymer-supported PTCs, can be easily recovered and reused. Most importantly, however, is that the total weight of multi-site PTC required in these reactions is less compared to related single-site PTCs. This is a decided advantage for the polymer-supported PTCs II-IV where the total weight of backbone polymer carrying an equivalent PTC activity was less than that of polymer-supported PTC I. In addition, as indicated in Table II, multi-site PTCs offer the potential for substrate-specific design in multi-leaving group substitution reactions. For example, while the differences in PTC activity among the single-site PTC tetrabutylphosphonium bromide (TBPB) and the TBPB analogs VI, IX, X and XI may be due in part to differences in lipophilicity, there also appears to be a substrate reaction site - PTC site effect for this disubstitution reaction. That is, TBPB and the 1,3(VI), 1,9(X) and 1,12(XI) di-site PTCs provide the same approximate level of efficiency while the 1,6(IX) di-site PTC is somewhat more active under the particular reaction conditions. The latter result indicates a more efficient transfer of nucleophile and suggests the possibility of a simultaneous, site-specific transfer in this disubstitution reaction. Additional investigations relative to this possibility are in progress.

Table I. Reaction of 1-Bromopentane with Various Nucleophiles under PTC Conditions[a]

$$\underline{n}\text{-}C_5H_{11}Br + K^+ Nu^- \xrightarrow{PTC} \underline{n}\text{-}C_5H_{11}Nu + K^+Br^-$$

PTC	YIELD(%), $\underline{n}$-C_5H_{11}Nu [b]			
	$Nu^- = PhO^-$	PhS^-	^-CN	$MeCOO^-$
NONE	33	7	8	18
NONE[c]	5	30	-	22
I[d]	81	83	79	70
II[e]	90->99[f]	90-98[f]	86-99	65-70
III[e]	85-95[f]	80-90[f]	75-90	65-70
IV[e]	>95	>95	75-80	60-70
V	85[f]	79[f]	80	74
VI	65	82	39	71
IX	85	92	76	59
X	79	90	37	73
XII[g]	66-83	86-93	68-77	60-73
XIII	70	62	19	94

[a] All reactions were run at 110°C by stirring a mixture of substrate, catalyst, water and nucleophile for 1.0 hr (PhO^-), 0.2 hr (PhS^-), 0.5 hr (^-CN) or 8 hr ($MeCOO^-$). The molar ratio of Nu/Substrate/PTC was 3/1/0.01 (PhO^-), 1.5/1/0.01 (PhS^-), 5/1/0.01 (^-CN) and 2/1/0.01 ($MeCOO^-$). Generally 33 mmoles of substrate, 0.3 mmoles of PTC, 25 ml of H_2O and the required amount of nucleophile were used. [b] Product isolated by extraction (Et_2O) and distilled. [c] DMSO used as co-solvent. [d] Yields from ref 13. [e] Yields represent range from three runs (catalyst recycled). [f] Yields from ref 18. [g] Yields represent range from three runs using different amount of catalyst for each run.

Table II. Reaction of Sodium Phenoxide with
          1,3-Dibromopropane under PTC Conditions

$Br(CH_2)_3Br \quad \xrightarrow[\text{Toluene, } 110°C, 1 \text{ hr}]{\text{PhONa, PTC}} PhO(CH_2)_3OPh$ Toluene,

PTC^a	Yield of Product (%)
None	0
TBPB [$\underline{n}$-Bu$_4$P$^+$Br$^-$]	45
VI [(CH$_2$)$_3$($\underline{n}$-Bu$_3$P$^+$Br$^-$)$_2$]	46
IX [(CH$_2$)$_6$($\underline{n}$-Bu$_3$P$^+$Br$^-$)$_2$]	65
X [(CH$_2$)$_9$($\underline{n}$-Bu$_3$P$^+$Br$^-$)$_2$]	47
XI [(CH$_2$)$_{12}$($\underline{n}$-Bu$_3$P$^+$Br$^-$)$_2$]	53

^{a}30 mole percent equivalent compared to substrate.

PTC Activity in a Weak Nucleophile SnAr Reaction

We have recently reported (26) several synthetic studies of weak nucleophile SnAr reactions. In the latter cases (26f-j), new synthetic methodology was reported for the direct introduction of fluoroalkoxy groups into a variety of aromatic systems. These reports represent synthetically useful procedures for obtaining some otherwise inaccessible fluoroalkoxy materials but, unfortunately, they require the use of a dipolar, aprotic solvent (usually hexamethylphosphoramide, HMPA) and, in some cases, elevated temperatures. However, because of their diverse and important applications (27), the syntheses of these and other organofluoro compounds continue to be of interest. For example, two recent reports of useful fluoroalkoxy materials include the insecticide activity exhibited by fluoroalkoxy substituted 1,3,4-oxadiazoles (28) and the control of cardiac arrythmia activity displayed by N-(Piperidylalkyl)fluoroalkoxybenzamides (29). While the efficacy of these new, biologically active materials has been attributed (28,29) primarily to the presence of the fluoroalkoxy group, it is important to note that the need for more efficient syntheses of these and related materials still exists. For example, the syntheses of both the above materials involve multi-step, moderate yield processes in which the fluoroalkoxy group is introduced in the initial step of the sequence via reaction of a nucleophilic phenolate derivative with either a fluoroalkene (30) or a triflate (31), respectively. Thus, the simplicity and convenience of the direct, SnAr introduction of fluoroalkoxy groups is an attractive alternative if conditions milder than those offered by elevated temperatures and HMPA as solvent can be established. A possible set of milder conditions for the latter reaction involves the use of phase-transfer catalysis.

While the greatest percentage of PTC-aided anionic substitutions involve non-aromatic systems (7-10), a number of liquid-liquid and solid-liquid, PTC-aided SnAr reactions have been reported (32-38). These reports involve a variety of substrates [unactivated (32,33), slightly activated (34), activated (35-37), and transition metal complexed (32,38)], nucleophiles [OMe (32,38), CN (33), SR (34), SCN (35), SO_3^{2-} (36), OR (37)] and PTCs [crown ethers (32-38), quats (33-35,38) and tertiary amine salts (36,37)]. In an attempt to extend this methodology to a weak nucleophile SnAr reaction, we have reported (39) a limited study of the effect of some well-known, single-site, neutral and ionic PTCs on the fluoroalkoxylation of a haloaromatic and a haloheteroaromatic substrate. We report here an extended study of the effect of single-site PTCs as well as that of some multi-site PTCs on aromatic fluoroalkoxylation. The results of these studies are shown in Tables III-V.

As indicated in Table III, neither 4-chloronitrobenzene nor 2-chloro-4-methylquinoline reacts with the 2,2,2-trifluoroethoxide ion in toluene at 25°C or at solvent reflux in the absence of catalyst. On the other hand, several different crown ethers and a range of poly(ethylene glycols) [PEGs of average molecular weight 300 to 14,000] catalyze this weak nucleophile SnAr reaction. The optimum conditions for these neutral PTCs are provided by use of PEG-8000 at 5 mole percent equivalent compared to substrate. As indicated in Figures 1 and 2, the relative orders of reactivity with respect to both the position of a leaving group and the nature of the leaving group follow the same pattern for the PEG-8000 mediated reaction as those reported previously by us (26i) and by others (40) for non-PTC mediated conditions. That is, position reactivity (Figure 1) is clearly o > p >> m as indicated by the complete conversion of the o-isomer during the first two hours of reaction and the absence of reaction of the m-isomer during the same time period. However, under PTC conditons, position reactivity is more selective (absence of reaction of the meta-isomer) whereas, under dipolar, aprotic solvent conditions, reaction of the meta-isomer is slower compared to the ortho- and para-isomers but still synthetically useful (26g,26i).

As shown in Table IV, a number of well-known, single-site, ionic PTCs also catalyze the aromatic fluoroalkoxylation reaction, albeit at somewhat higher molar ratios than that required for the optimum conditions using PEG-8000. The most effective of these ionic materials was tetrabutylphosphonium bromide (TBPB) which is in agreement with a study by Brunelle (34c) on the PTC-mediated reaction of thiolates with polychlorobenzenes. It is interesting to note that solid-liquid conditions generally provided better conversions for the ionic PTCs than liquid-liquid conditions and that water

Table III. Reaction of 4-Chloronitrobenzene with 2,2,2-Trifluoroethoxide Ion under Neutral PTC Conditions[a]

$$4\text{-Cl-Ph-NO}_2 + CF_3CH_2OH \xrightarrow[\text{PhMe, 24h, 110°C}]{\text{PTC}} 4\text{-}CF_3CH_2O\text{-Ph-NO}_2$$

(1 eq)   (1.5 eq)   NaOH(s, 1.1 eq)

PTC (eq)	CONVERSION (% BY GC)	YIELD (%, CRUDE)
–[b]	0	0
18-C-6[c] (0.10)[d]	43	57
18-C-6 (0.28)[d,e]	58	71
18-C-6 (0.28)[e]	64	51
18-C-6[f] (0.44)[e]	98	75
15-C-5[f] (0.44)[e]	60	66
DB18-C-6[g] (0.44)[e]	57	74
PEG-300[h] (0.44)	13	56
PEG-600 (0.44)	37	44
PEG-1000 (0.44)	40	54
PEG-1500 (0.44)	50	43
PEG-3400 (0.44)	61	36
PEG-8000 (0.44)[e]	76	52
PEG-14000 (0.44)	63	47
PEG-8000 (0.66)[i]	78	41
PEG-8000 (0.30)[i]	97	44
PEG-8000 (0.20)[i]	92	62
PEG-8000 (0.20)[j]	52	60
PEG-8000 (0.10)[i]	97	59[j]
PEG-8000 (0.05)[i,e]	97	78[j]
WITH 2-CHLORO-4-METHYLQUINOLINE AS SUBSTRATE		
PEG-8000 (0.05)[i,e]	100	>98[k]

[a] All reactions were run under a nitrogen atmosphere. Initial product isolation was accomplished by aqueous extraction of the reaction mixture followed by concentration of the organic phase. Azeotropic removal of water during the course of the reaction did not affect the indicated results. [b] At 25°C and at 110°C. [c] 18-Crown-6. [d] Base used was NaH(1.1 equiv). [e] Data from ref 39. [f] 15-Crown-5. [g] Dibenzo-18-Crown-6. [h] PEG = Poly(ethyleneglycol) of molecular weight indicated. [i] 1.7 equiv of NaOH and 2.0 equiv of CF_3CH_2OH were used. [j] Isolated yield = 41%. [k] Isolated yield = 74%.

Table IV. Reaction of 4-Chloronitrobenzene with
2,2,2-Trifluoroethoxide Ion under Single-Site,
Ionic PTC Conditions[a]

$$4\text{-Cl-Ph-No}_2 + CF_3CH_2OH \xrightarrow[\text{NaOH(s, 1.2 eq)}]{\text{PTC}} 4\text{-}CF_3CH_2\text{O-PH-NO}_2$$
(1 eq)   (2 eq)   PhMe, 24h, 110°C

PTC (eq)		CONVERSION (% BY GC)	YIELD (%.CRUDE)
BTMAC[b]	(0.1)	16	14
BTEAC[c]	(0.1)[d]	9	2
BTEAC	(0.1)	15	10
MTCAC[e]	(0.1)	9	5
TBAHS[f]	(0.1)	26	16
AD464[g]	(0.1)[h]	25	21
AD464	(0.1)	11	9
TBPB[i]	(0.1)[d]	49	32
TBPB	(0.1)	30	19
TBPB	(0.3)	99	84
TBPB	(0.3)[j]	88	68
TBPB	(0.5)	95	78
WITH 2-CHLORO-4-METHYLQUINOLINE AS SUBSTRATE			
TBPB	(0.3)	98	75[k]

[a] Unless otherwise indicated, all reactions were run under solid-liquid conditions in a nitrogen atmosphere and water was azeotropically removed during the course of the reaction. Initial product isolation was accomplished by rapid filtration of the reaction mixture through a silica gel column followed by concentration of the organic phase. Data for BTEAC liquid-liquid conditions and for TBAHS from ref 39. [b] Benzyltrimethylammonium chloride. [c] Benzyltriethylammonium chloride. [d] Liquid-liquid conditions: catalyst added as a 40% (w/w) aqueous solution. [g] Adogen 464 (Ashland Chemical Company): Methyltrialkyl-(C_{8-10}) ammonium chloride. [h] Liquid-liquid conditions: water added directly to reaction mixture. [i] Tetrabutylphosphonium bromide. [j] Water was not removed azeotropically during the course of the reaction. [k] Isolated yield from ref 39.

Table V. Reaction of 4-Chloronitrobenzene with
2,2,2-Trifluoroethoxide Ion under Multi-Site,
Ionic PTC Conditions[a]

$$4\text{-Cl-Ph-NO}_2 + CF_3CH_2OH \xrightarrow[\text{PhMe,2h,110°C}]{\text{PTC(0.3eq)} \atop \text{NaOH(s,1.2eq)}} 4\text{-}CF_3CH_2O\text{-Ph-NO}_2$$

(1 eq)   (2 eq)

PTC	YIELD (%,ISOLATED)
VI	48
VII	58 (58)[b]
VIII	70
IX	70
X	65 (65)[b]
XI	71
TBPB	12[c] (65 at 24hr)[d]
TBPB	trace[b]
PEG-8000	-[e] (41 at 24hr)[f]

[a] See footnote a, Table IV. [b] Water was not removed azeotropically during the course of the reaction.
[c] Corresponds to approximately 50% conversion at 2hr.
[d] Corresponds to 99% conversion at 24hr (See Table IV).
[e] Approximately 50% conversion at 2hr (See Figure 1).
[f] Corresponds to 97% conversion at 24hr (See Table III).

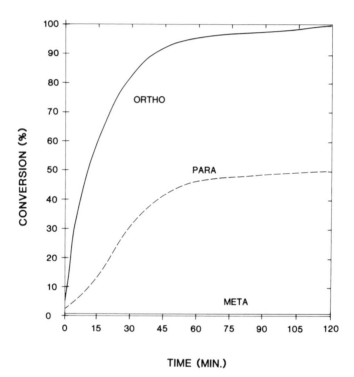

Figure 1. Best-fitted conversion (%)/time (min) curves for the PEG-8000 mediated fluoroalkoxylation of o-, m- and p-chloronitrobenzenes

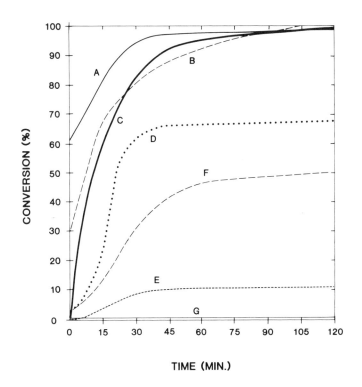

Figure 2. Best-fitted conversion (%)/time (min) curves for the PEG-8000 mediated fluoroalkoxylation of substituted nitrobenzenes A = o-NO_2, B = o-F, C = o-Cl, D = o-Br, E = o-I, F = p-Cl, G = m-Cl

appears to inhibit the ionic PTC-catalyzed reactions but not the PEG-catalyzed reactions. In fact, it was necessary to remove water azeotropically from the reactions employing ionic PTCs in order to achieve optimum conversion. Such behavior seems to be unusual in light of the many reported (9) applications of ionic PTCs to liquid-liquid transfers but may be indicative of "omega-phase catalysis" as mentioned in the following discussion and as described in detail by Liotta (41) elsewhere in this volume.

The data reported in Table V provide a summary of the reaction of 2,2,2-trifluoroethoxide ion with 4-chloronitrobenzene in the presence of the di-site phosphonium PTCs VI-XI. As indicated, optimum conditions for the di-site PTCs are obtained after 2 hr of reaction. Under such conditions, the various di-site PTCs are more effective than either TBPB (at the same molar ratio) or PEG-8000.

In addition, in contrast to TBPB, the effectiveness of the di-site PTCs are not inhibited by the presence of water. The latter observation may explain why the single-site, ionic PTCs are generally more effective under solid-liquid conditions than under liquid-liquid conditions and, in contrast, why the di-site, ionic PTCs are more effective. That is, for a PTC-mediated reaction carried out in the presence of gradually increasing amounts of water, Liotta (41) has presented evidence which indicates that the PTC moves from the organic phase onto the surface of the salt nucleophile. Liotta suggests that a new phase is thus formed (i.e., the "omega-phase") where reaction then occurs on the surface of the salt nucleophile. Under normal liquid-liquid PTC conditions, where other than trace amounts of water are involved, there is a subsequent decrease in reactivity. In comparing the reactivities of the single-site, ionic PTCs to those of the multi-site, ionic PTCs, "omega-phase" formation may well be an inhibiting factor in the former case but appears to be of little significance in the latter case.

Summary

Thus, PTC methodology can be effectively applied to a weak nucleophile SnAr reaction under a variety of PTC and reaction conditions. This may prove to be particularly useful for fluoroalkoxylation. That is, we have recently demonstrated (26h) that 2,2,2-trifluoroethoxide ion reacts under HMPA solvent conditions at or near room temperature with an extended range of substrates which contain a more active leaving group (nitro) than chloro. The former reaction occurs usefully even for substrates containing the weakly activating amido functionality. Similiar reactions may be possible for these substrates under PTC conditions.

Acknowledgment

We are grateful to the Petroleum Research Fund, administered by the American Chemical Society, and the Robert A. Welch Foundation for partial support of this work. We thank the following co-workers for their experimental assistance and expertise: Ronald Wysocki, Sherre Young, Jeffery Turcot, Charlene Ohlman, Russell Leonard, Jeffery Dodge, Jill Garrison, Craig Hughes, Joseph Coury, Martin Moebus and Der-Lun Chu.

Literature Cited

1. Starks, C. M.   J. Am. Chem. Soc. 1971, 93, 195.
2. Starks, C. M.   ibid. 1973, 95, 3613.
3. Sjoberg, K.   Aldrichimica Acta 1980, 13, 55.
4. Jones, A. R.   ibid. 1976, 9, 35.
5. Gokel, G. W.; Weber, W. P.   J. Chem. Educ. 1978, 55, 350 and 429.
6. Reuben, B.; Sjoberg, K.   Chemtech 1981, 315.
7. Dehmlow, E. V.; Dehmlow, S.   "Phase Transfer Catalysis," Verlag Chemie, Weinheim and Deerfield Beach, Florida, 1980.
8. Keller, W. E.; Ed., "Compendium of Phase-Transfer Reactions and Related Synthetic Methods," Fluka AG, CH-9470 Buchs, Switzerland, 1979.
9. Liotta, C.; Starks, C. M.   "Phase Transfer Catalysis," Academic Press, New York, 1977.
10. Weber, W. P.; Gokel, G. W.   "Phase Transfer Catalysis in Organic Synthesis," Springer-Verlag, New York, 1977.
11. Gokel, G. W.; Jarvis, B. B.   "Medium Effects in Organic Synthesis," ACS Audio Course, 1983.
12. Mathur, N. K.; Narang, C. K.; Williams, R. E. "Polymers as Aids in Organic Chemistry," Academic Press, New York, 1980.
13. (a) Chiles, M. S.; Reeves, P. C.   Tetrahedron Letters 1979, 3367; (b) Chiles, M. S.; Jackson, D. D.; Reeves, P. C.   J. Org. Chem. 1980, 45, 2915.
14. Regen, S. L.; Bolikal, D.; Barcelon, D.   J. Org. Chem. 1981, 46, 2511.
15. Fukunishi, K.; Czech, B.; Regen, S. L.   ibid. 1981, 46, 1218.
16. Montanari, F.; Tundo, P.   ibid. 1981, 46, 2125.
17. Regen, S. L.   Angew. Chem. Int. Ed. Engl. 1979, 18, 421.
18. Idoux, J. P.; Wysocki, R.; Young, S.; Turcot, J.; Ohlman, C.; Leonard, R.   Synth. Commun. 1983, 13, 139.
19. (a) Brown, J.; Jenkins, J.   J. Chem. Soc., Chem. Commun. 1976, 458; (b) Cinouini, M.; Colonna, S.; Molinari, H.; Montanari, F.; Tundo, P.   ibid. 1976, 394; (c) Molinari, H.; Montanari, F.; Tundo, P.

ibid. 1977, 639; (d) Molinari, H.; Montanari, F.; Quinci, S.; Tundo, P. J. Am. Chem. Soc. 1979, 101, 3920.
20. (a) Tomoi, M.; Ogama, E.; Hosokawa, Y.; Kaiuchi, H. J. Polym. Sci., Polym. Chem. Ed. 1982, 20, 3015; (b) Tomoi, M.; Hosokawa, Y.; Kaiuchi, H. ibid. 1984, 22, 1243.
21. Cinquini, M.; Montanari, F.; Tundo, P. J. Chem. Soc. Chem. Commun. 1975, 393.
22. Landini, D.; Rolla, F. Synthesis 1974, 565.
23. McKillop, A.; Fiaud, J. C.; Hug, R. P. Tetrahedron 1974, 30, 1379.
24. Herriott, A.; Picker, D. Synthesis 1975, 447.
25. Tundo, P. Synthesis 1978, 315.
26. Idoux, J. P.; Gupton, J. T.; et al.: (a) Synth. Commun. 1985, 15, 43; (b) ibid. 1984, 14, 1001; (c) ibid. 1983, 13, 1083; (d) ibid. 1982, 12, 907; (e) J. Org. Chem. 1983, 48, 2933; (f) Synth. Commun. 1984, 14, 621; (g) ibid. 1982, 12, 695; (h) J. Org. Chem. 1985, 50, 1876; (i) ibid. 1983, 48, 3771; (j) Can. J. Chem. 1985, 63, 3037.
27. Gerstenberger, M.; Haas, A. Angew. Chem. Int. Ed. Engl. 1981, 20, 647.
28. (a) Arrington, J.; Wade, L., U.S. Patent 4,215,129, July 29, 1985; (b) Idoux, J. P.; Cunningham, G. N.; Gupton, J. T., unpublished results reported in part at the 1984 Southeast Regional ACS Meeting.
29. Chem. & Engr. News 1984, September 10, pp. 32-38.
30. Idoux, J. P.; Gibbs-Rein, K., unpublished results.
31. Banitt, E. H.; Brown, W. R.; Coyne, W. E.; Schmid, J. R. J. Med. Chem. 1977, 20, 821.
32. Fukui, M.; Endo, Y.; Oishi, T. Chem. Pharm. Bull. 1980, 28, 3639.
33. Cassar, L.; Foa, M.; Montanari, F.; Marinelli, G. P. J. Organomet. Chem. 1979, 173, 335.
34. (a) Landini, D.; Montanari, F.; Rolla, F. J. Org. Chem. 1983, 48, 604; (b) Reeves, W.; Bothwell, T.; Rudis, J.; McClusky, J.; Synth. Commun. 1982, 12, 1071; (c) Brunelle, D. J. Org. Chem. 1984, 49, 1309.
35. Reeves, W. P.; Simmons, A.; Keller, K. Synth. Commun. 1980, 10, 633.
36. Bunnett, J. F.; Gisler, M.; Zollinger, H. Helv. Chim. Acta 1982, 65, 63.
37. Paradisi, C.; Quintily, U.; Scorrano, G. J. Org. Chem. 1983, 48, 3022.
38. Montanari, F.; Pelosi, M.; Rolla, F. Chem. Ind. (London) 1982, 412.
39. Gupton, J. T.; Coury, J.; Moebus, M.; Idoux, J. P. Synth. Commun. 1985, 15, 431.
40. March, J. "Advanced Organic Chemistry"; 3rd ed.; McGraw-Hill: New York, 1985; Chapter 10.
41. Liotta, C. L. "Omega-Phase Catalysis," this volume.

RECEIVED July 26, 1986

INDEXES

Author Index

Alper, Howard, 8
Anelli, Pier Lucio, 54
Arnold, K., 24
Bhattacharya, A., 67
Black, Elzie D., 15
Brown, Keith C., 82
Brunelle, Daniel J., 38
Burgess, Edward M., 15
Cleary, T., 24
Dimotsis, George, 128
Dolling, U.-H., 67
Fair, Barbara E., 15
Friese, R., 24
Gatto, V., 24
Gerbi, Diana J., 128
Gokel, George W., 24
Goli, D., 24
Grabowski, E. J. J., 67
Gupton, John T., 169
Hanlon, C., 24
Heilmann, Steven M., 116
Hughes, D. L., 67
Idoux, John P., 169
Karady, S., 67
Kellman, Raymond, 128
Kim, M., 24
Krepski, Larry R., 116

Landini, Dario, 54
Lee, Donald G., 82
Lee, Eric J., 82
Liotta, Charles L., 15
Maia, Angelamaria, 54
Miller, S., 24
Montanari, Fernando, 54
Nicholas, Paul P., 155
Ouchi, M., 24
Percec, Virgil, 96
Posey, I., 24
Quici, Silvio, 54
Rasmussen, Jerald K., 116
Ray, Charles C., 15
Ryan, K. M., 67
Sandler, A., 24
Sawicki, Robert A., 143
Smith, Howell K. II, 116
Starks, Charles M., 1
Viscariello, A., 24
Weinstock, L. M., 67
White, B., 24
Williams, Janet C., 128
Williams, Robert F., 128
Wolfe, J., 24
Yoo, H., 24

Subject Index

A

Acetone, use in permanganate oxidations, 92-93
Acrylonitrile, polymerization, 120
Activity of phase-transfer catalysts
S_N2 reactions, 170-175
weak-nucleophile S_NAr reactions, 175-182
Acyltetracarbonyl cobalt compound, cleavage in the carboxyalkylation of alkyl halides, 150
Addition reactions, Michael, catalytic asymmetric, 69,70f
Alcohol sidearms, bibracchial lariat ethers, 34
Alkenes
oxidation, 84-85
permanganate oxidation, 82-93

Alkylating agent, effect on devulcanization, 161
Alkylations
asymmetric, chiral phase-transfer catalysis, 67-79
pyridines, 44
3-(Alkylthio)cyclohexenes, 166-167
Alkynes
oxidation, 85-86
permanganate oxidation, 82-93
Allyl halides, carbonylation to unsaturated acids, 13
Alumina, anionically activated, use in phase-transfer catalysis, 147-150
Ammonium, cation binding by lariat ethers, 30,32
Ammonium persulfate, use in free radical polymerization, 118

Ammonium salts
  increasing chemical resistance
    via resonance stabilization, 43
  increasing stability via steric
    hindrance, 43
Anion(s)
  activated by lipophilic
    macropolycyclic ligands,
    phase-transfer catalysis, 58-60
  hydration and reactivity,
    phase-transfer catalysis, 55-57
Anion chemistry, organometallic,
  triphase catalysis, 143-153
Anionically activated alumina, use in
  phase-transfer catalysis, 147-150
Annulation, chiral Robinson, 79,80f
Aqueous solution, ion
  extraction, 96-99
Aromatic hydrocarbons, hydrogenation
  by phase-transfer catalysis, 10
Aromatic substitution in condensation
  polymerization, catalyzed by
  solid-liquid phase
  transfer, 128-141
Aryl displacement reactions,
  phase-transfer catalysis, 51
Asymmetric alkylations via chiral
  phase-transfer catalysis, 67-79

B

Base
  effect of composition on
    carboxymethylation
    reactions, 148-149
  effect of concentration on rate of
    reactions promoted by alkali
    hydroxides, 56-57
  extractable, versus
    carboxyalkylation activity, 150
Benzyl bromide
  carbonylation, 146-147
  carboxymethylation, 147-148
Benzyl chloride, effect on
  devulcanization, 161
Benzyl halides
  carbonylation, 146-147
  reaction with potassium
    cyanide, 15-22
Benzyltriethylammonium chloride, as a
  phase-transfer catalyst, 147
Bibracchial lariat ethers
  synthesis, 34
  use as phase-transfer
    catalysts, 32,34
Bimetallic phase-transfer reaction, 12
Binding process, description, 25
Bisammonium salt phase-transfer
  catalysts, 45-48
Bisaryl substrates, polycondensation
  reactions, 130,133

Bis-(4-dihexylaminopyridinium)decane
  dibromide, preparation, 51
Bis-1,4-(disubstituted)perfluoro-
  benzene, preparation, 138,140
Bis-4,4'-(disubstituted)perfluoro-
  biphenyl, preparation, 138,140
Bishaloaromatics, polymerization with
  bisphenols and
  bisthiophenol, 130,132
Bis-(4-nitrophenyl) ether of
  bisphenol-A, preparation, 48,49f
Bisnucleophiles, polymerization, 133
Bisphenol(s), polymerization, 130,132
Bisphenol-A, polymerization, 133
Bisphenol-A and hexafluorobenzene,
  effect of water on
  polymerization, 135-136,139f
Bisquaternary salts, as catalysts, 4
Bisthiophenol, polymerization, 130,132
Branching, polymerizations with
  hexafluorobenzene, 130
1-Bromobutane, reaction with potassium
  acetate, 145
1-Bromopentane, reaction with
  nucleophiles, 173,174t
Butenolides, formation in
  phase-transfer reactions, 12
Butyl acrylate,
  polymerization, 118,119f
4-tert-Butylthiophenol, reaction with
  hexafluorobenzene, 129

C

Carbon-pivot lariat ethers, 29
Carbonylation
  benzyl halides, 146-147
  use of phase-transfer
    catalysis, 11-13
Carboxyalkylation, propylene
  oxide, 151-152
Carboxyalkylation activity, versus
  extractable base, 150
Carboxymethylation, benzyl
  bromide, 147-148
Catalysis
  aromatic substitution in
    condensation
    polymerization, 128-141
  4-cresyl-4'-nitrophenyl
    ether formation, 39-41
  phase transfer
    chiral, 67-79
    mechanism, 15-23
    overview, 1-5
    use of metal complexes, 8-13
    triphase, organometallic anion
      chemistry, 143-153
  See also Phase-transfer catalysis
Catalysts
  cation-binding properties, 24-35

INDEX

189

Catalysts--Continued
 decomposition
  pathways, 74f
 scission of polysulfide
  cross-links, 156-161
 effect of amount on reaction
  rate, 45
 effect on hexafluorobenzene
  polymerization, 133-134
 effect on perfluorobiphenyl
  polymerization, 133,135
 efficiency coefficients, 62-64
 half-lives in the presence of sodium
  phenoxide, 44t
 improvements, 3-5
 multisite phase transfer, 169-182
 rate dependence, 28t
 soluble and polymer supported, 54-65
 stable, for phase transfer at
  elevated temperatures, 38-51
 See also Phase-transfer catalysts,
  Polymer-supported phase-transfer
  catalysts
Catalyst-indanone ratio, effect on
 rate and selectivity of chiral
 methylation, 75t
Catalytic activity of sandwich
 ligands, nucleophilic
 substitutions, 59t
Catalytic cycle, carboxyalkylation of
 alkyl halides, 150-151
Cation(s), methanol solutions
 containing crown ethers, stability
 constants, 26t
Cation-binding properties, polyether
 species, 24-35
Cation-binding selectivities,
 crown ethers, 26
Cation complexes of lipophilic
 multidentate ligands, as catalysts
 in two-phase systems, 60
Chain-ended functional polymers,
 polyetherification, 107
Chemical resistance of ammonium salts,
 use of resonance stabilization to
 increase, 43
Chiral phase-transfer catalysis
 asymmetric alkylations, 67-79
 catalysts, 3-4
 reaction profile, 76f
Chiral Robinson annulation, 79,80f
Chloronitrobenzenes,
 fluoroalkoxylation, 176,180-181f
Cinchonidine catalyst,
 interactions, 79,80f
Cinchonine catalyst,
 interactions, 79,80f
Cobalt carbonyl, as a catalyst in
 carbonylation reactions, 11-12
Cobalt tetracarbonyl anion catalyst,
 use in carbonylation of benzyl
 halides, 146

Complex(es)
 metal, phase-transfer
  catalysis, 8-13
 solid state, structures, 32
Complexation ability, crown
 ethers, 26,58
Condensation polymerization, catalyzed
 by solid-liquid phase
 transfer, 128-141
Conventional step polymerization,
 general discussion, 97-98
Copolymers, sequential, synthesis, 107
4-Cresyl-4'-nitrophenyl ether,
 formation, 39-41
Cross-links
 density determinations, 164-165
 polymerizations using
  hexafluorobenzene, 130
 polysulfide in rubber particles,
  scission through phase-transfer
  catalysis, 155-167
18-Crown-6
 concentration in toluene, effect of
  added water, 21t,22f
 methyl methacrylate
  polymerization, 121
 reaction of benzyl halides with
  potassium cyanide, 15-22
Crown ethers
 catalytic activity, 61-63
 catalytic effect in polymerization
  reactions, 135
 complexation ability, 58
 polymer supported, 62-63
 use as phase-transfer
  catalysts, 25-27,118,119f
Cryptands
 catalytic activity, 61-64
 polymer supported, 63-64
Cryptates, phase-transfer
 catalysis, 56,58-60
Crystal structures, lariat ethers, 32
Cyclic ethers, catalytic
 effectiveness, 4
Cyclic manganate(V) diesters, 86-87
Cyclodextrins, as phase-transfer
 catalysts, 11,117-119
2-Cyclohexen-1-thiol, preparation, 166

D

Decomposition pathways, catalysts, 74f
De-2-cyclohexen-1-yl disulfide,
 preparation, 165-166
Demethylation, inhibition, 161
Devulcanization
 in the presence of benzyl chloride
  and methyl chloride, 161-163
 of rubber peelings, 164

Dialkylaminopyridinium salts, as
  stable phase-transfer
  catalysts, 4,41-48
1,3-Dibromopropane, reaction with
  sodium phenoxide, 173,175t
Di-2-cyclohexen-1-yl disulfide
  disulfide bond
    scission, 158,160f,162f
  phase-transfer-catalyzed
    decomposition, 167
α,ω-Di(electrophilic) aromatic
  polyether sulfones, synthesis, 103
α,ω-Di(electrophilic) oligomer, H-1
  NMR spectrum, 104f
4,4'-Dihydroxybiphenyl polyethers and
  copolyethers, synthesis, 112
4,4'-Dimercaptodiphenyl sulfide,
  polymerization with fluorinated
  bisaryls, 133
α,ω-Di(2-(p-phenoxy)-2-oxazoline)
  oligomers
  H-1 NMR spectrum, 105f
  synthesis and reactions, 101-102
Disulfides, dismutation
  reaction, 156,157

E

Efficiency coefficients of
  catalysts, 62-64
Efficient asymmetric alkylations via
  chiral phase-transfer
  catalysis, 67-79
Electrophilic attack, permanganate
  oxidations, 90
Electrophilic groups,
  phase-transfer-catalyzed
  polymerizations, 98-99
Epoxidations, catalytic
  asymmetric, 69,70f
Ester and ether selectivity,
  carboxymethylation of benzyl
  bromide, 148-149
Ether sidearms, bibracchial lariat
  ethers, 34
Ethyl acrylate, polymerization, 120
N-(2-Ethylhexyl)-4-dimethylamino-
  pyridinium mesylate,
  preparation, 51
Extraction constants
  transfer of permanganate from water
    into methylene chloride, 83t
  two-phase system containing a
    hydrophobic salt dissolved in
    water, 96-97
Extraction of ions from aqueous
  solution, 96-99

F

Fluorinated bisaryls, polymerizations
  with bisphenol-A and
  4,4'-dimercaptodiphenyl
  sulfide, 133
Fluoroalkoxylation,
  nitrobenzenes, 176,180-181f
Free radical pathway,
  phase-transfer-catalyzed
  polycondensations, 138-139
Free radical polymerizations, phase
  transfer, mechanistic
  aspects, 116-126
Functional polymers containing cyclic
  imino ethers, 99-107

G

Gel permeation chromatograms, triblock
  copolymer, 111f

H

H-1 NMR spectra, polymer synthesis by
  phase-transfer
  catalysis, 104-105f,108f,110-11f
Half-lives of catalysts in the
  presence of sodium phenoxide, 44t
Halides and pseudohalides
  nucleophilic reactivity, 56
  reactivity scale, 56
Hammett plots
  chiral phase-transfer
    methylation, 73f
  permanganate oxidations, 88-89f
Heating scans, poly(2,6-dimethyl-
  1,4-phenylene oxide), 108f
Heterocyclic compounds, 10
Hexadecylpyridinium chloride, use in
  methyl methacrylate
  polymerization, 120
Hexadecyltrimethylammonium bromide,
  use in acrylonitrile
  polymerization, 120
Hexadecyltrimethylammonium persulfate,
  use in acrylonitrile
  polymerization, 120
Hexafluorobenzene
  polycondensation reactions, 129
  polymerization
    effect of solvent and
      catalyst, 133-134
    effect of water, 135-136,139f
    with bisphenols and
      bisthiophenol, 130,132
Hofmann elimination
  phase-transfer catalysis, 56-57
  procedure, 165

INDEX

Hofmann reaction, catalyst decomposition, 156-159
Hole-size relationship of crown ethers, evidence concerning, 25-27
Homologation, oxidative, method, 12
Hydration and reactivity of anions, phase-transfer catalysis, 55-57
Hydrocarbons, aromatic, hydrogenation by phase-transfer catalysis, 10
Hydrogen substitution, benzene nucleus as a function of charge type, 136
Hydrogenation, aromatic hydrocarbons and heterocyclic compounds, 10
2-(p-Hydroxyphenyl)-2-oxazoline, synthesis and reactions, 100

I

Imino ethers, cyclic, functional polymers containing, 99-107
Immobilized polyalkylene glycols, catalytic ability, 144-145
Indanone
  chiral phase-transfer methylation, 69,71f
  interfacial deprotonation, 75-77
Indanone anion, ion pairing with benzylcinchoninium cation, 69-72
Indanone-catalyst ratio, effect on chiral methylation, 75t
Initiation of phase-transfer free radical polymerizations, 117,119
Interfacial deprotonation, indanone, 75-77
Intramolecularity in lariat ether complexes, 30,32
Ion extraction from aqueous solution, 96-99
Ion pairing between cations and anions, 69-72
4-Isopropylphenol, reaction with hexafluorobenzene, 129
Isotope effects, permanganate oxidations, 92

K

Ketones, approaches for asymmetric alkylations, 67
Kinetics
  phase-transfer free radical polymerizations, 123-124,125
  reaction of benzyl halides with potassium cyanide, 15-22

L

Lactones, formation by using phase-transfer catalysis, 13

Lariat ethers
  intramolecularity, 30,32
  use as phase-transfer catalysts, 29
  X-ray crystal structure analysis, 32,33f
Lipophilic macropolycyclic ligands as anion activators, 58-60
Liquid crystalline main-chain polymers, 107,112-113
Liquid-solid phase-transfer catalysis, aromatic substitution in condensation polymerization, 128-141
Long-chain linear polyethers, use as phase-transfer agents, 84,85t

M

Manganese carbonyl complexes, use in phase-transfer catalysis, 13
Mechanisms
  chiral phase-transfer alkylations, 72-79
  permanganate oxidations, 86-93
  phase-transfer catalysis, 15-23
Mercaptans, phase-transfer-catalyzed reactions, 9
Metal complexes, phase-transfer catalysis, 8-13
Metal oxide surfaces, immobilized poly(alkylene glycols), 145
Metallocyclooxetane, formation during permanganate oxidations, 90
Methyl chloride, effect on devulcanization, 161
Methyl 3-hydroxybutrate, production, 151-152
Methyl methacrylate, polymerization, 120-125
Methyl phenylacetate, production, 147-150
Methylations
  chiral phase-transfer catalysis, 69-72
  racemic and chiral, order in catalyst, 77-79
  rate, initial versus delayed catalyst charge, 77,78f
Methylene chloride, use as solvent for permanganate oxidations, 92-93
Methyltricaprylylammonium chloride, use in methyl methacrylate polymerization, 121-123
Model cross-linked polybutadiene, scission of polysulfide cross-links, 156,159f
Multisite phase-transfer catalysts, 4,169-182

N

Neopentylation, phase-transfer
 catalysis, 48,51
N-Neopentyl-4-dihexylaminopyridinium
 bromide, preparation, 51
Neutral phase-transfer catalysis,
 reaction of 4-chloronitrobenzene
 with 2,2,2-trifluoroethoxide
 ion, 176,177t
Nitro compounds, reduction by
 phase-transfer catalysis, 8
Nitrobenzenes,
 fluoroalkoxylation, 176,180-181f
Nitrogen-pivot lariat ethers, 29-30
4-Nitrophenyl phenyl ether,
 preparation, 45,47f
Nonpolymeric, multisite,
 phase-transfer catalysts, 173
Nuclear magnetic resonance assays,
 chiral phase-transfer
 methylation, 74f
Nucleophiles
 polymerization reactions, 133
 reactions with
  1-bromopentane, 173,174t
Nucleophilic aromatic displacement
 reactions, 39
Nucleophilic aromatic substitution,
 general discussion, 129
Nucleophilic groups,
 phase-transfer-catalyzed
 polymerizations, 98-99
Nucleophilic reactivity, halides and
 pseudohalides, 56
Nucleophilic substitutions
 catalytic efficiency of crown
  ethers, 63t
 catalytic efficiency of
  cryptands, 64t
 catalytic efficiency of sandwich
  ligands, 59t

O

n-Octyl methanesulfonate, nucleophilic
 substitutions, 59t,63t,64t
Olefins, catalyzed oxidation, 11
Oligoethylene glycols, use as
 phase-transfer catalysts, 27-29
Omega phase, phase-transfer-catalyzed
 reactions, 19,23
Onium bisphenolates, ion extraction
 from aqueous solution, 98
Onium salts, use as phase-transfer
 catalysts, 24-25,163-164
Organometallic anion chemistry,
 triphase catalysis, 143-153
Oxidation
 alkenes and alkynes by
  permanganate, 82-93

Oxidation--Continued
 use of phase-transfer
  catalysts, 10-11
Oxidative homologation, method, 12

P

Palladium, as a catalyst in oxidation
 reactions, 10-11
Perfluorobenzophenone,
 polycondensation
 reactions, 130,133
Perfluorobiphenyl
 effect of solvent and catalyst on
  polymerization, 133,135
 polycondensation reactions, 130,133
Perfluorophenyl sulfide,
 polycondensation
 reactions, 130,133
Permanganate
 oxidation of alkenes and
  alkynes, 82-93
 transfer from an aqueous phase into
  methylene chloride, 83
Phase-transfer catalysis
 alkylation and neopentylation, 48,51
 applications, 2-3,144
 aromatic substitution in
  condensation
  polymerization, 128-141
 aryl displacement reactions, 51
 catalyst improvements, 3-5
 chiral, asymmetric
  alkylations, 67-79
 development, 5
 free radical
  polymerizations, 116-126
 mechanisms, 15-23,55,61
 overview, 1-5
 permanganate oxidation of alkenes
  and alkynes, 82-93
 polymer chemistry, 5
 polymer synthesis, 96-113
 scission of polysulfide cross-links
  in rubber particles, 155-167
 stable catalysts at elevated
  temperatures, 38-51
 use of metal complexes, 8-13
 See also Catalysis
Phase-transfer catalysts
 cation-binding properties, 24-35
 immobilized into a polymeric
  matrix, 60-64
 multisite, 169-182
 soluble and polymer supported, 54-65
 See also Catalysts, Polymer-supported
  phase-transfer catalysts
Phenoxides, nucleophilic aromatic
 substitution reactions, 39

INDEX

Phenylacetic acid, preparation, 146
Polarity of solvent, effect on
  polymerization rate, 121-122
Poly(alkylene glycols), catalytic
  ability, 144-145
Polybutadiene, scission of polysulfide
  cross-links, 156,159f
Polybutadiene rubbers
  containing disulfide cross-links,
    preparation, 165
  molecular weight
    distribution, 156,159f
Poly(2,6-dimethyl-1,4-phenylene oxide)
  H-1 NMR spectrum, 108f
  heating scans, 108f
  synthesis, 106
Polyether(s)
  bound to cations, as catalysts, 5
  bound to silica, as catalysts, 4
  cation-binding properties, 24-35
  linear, use as phase-transfer
    agents, 84,85t
  thermotropic, 107,112-113
Polyetherification
  chain-ended functional polymers, 107
  implications of ion extraction from
    aqueous solution, 96-99
Poly(ethylene glycols)
  catalysis of weak-nucleophile S_NAr
    reactions, 176,180-181f
  complexation of salts, 84
  use as phase-transfer
    catalysts, 10,27-29,39-41
Polymer(s), functional, containing
  cyclic imino ethers, 99-107
Polymer chemistry, phase-transfer
  catalysis, 5
Polymer-supported crown ethers,
  catalytic activity, 62-63
Polymer-supported cryptands, formation
  and catalytic activity, 63-64
Polymer-supported phase-transfer
  catalysts
  activity in S_N2 reactions, 170-175
  catalytic abilities, 172-173
  general groups, 29
  preparation, 170-172
  reactivity and application, 54-65
  See also Catalysts, Phase-transfer
    catalysts
Polymer-supported quaternary onium
  salts, formation and catalytic
  activity, 60
Polymeric matrix, immobilization of
  phase-transfer catalysts, 60-64
Polymerization
  acrylonitrile, 120
  bisphenol-A and 4,4'-dimercapto-
    diphenyl sulfide, 133
  bisphenol-A and hexafluoro-
    benzene, 135-136,139f

Polymerization--Continued
  bisphenols and
    bisthiophenol, 130,132
  butyl acrylate, 118,119f
  catalyzed by solid-liquid phase
    transfer, 128-141
  conventional step, general
    discussion, 97-98
  effect of water, 135-141
  ethyl acrylate, 120
  hexafluorobenzene, effect of solvent
    and catalyst, 133-134
  methyl methacrylate, 120-125
  perfluorobiphenyl, effect of solvent
    and catalyst, 133,135
  phase-transfer free radical, 116-126
  styrene, 121
  use of phase-transfer
    catalysis, 96-113
Polysulfide cross-links, scission in
  rubber particles, 155-167
Polyvinylpyrrolidone, as a stable
  phase-transfer catalyst, 39-41
Potassium cyanide, reaction with
  benzyl halides, 15-22
Potassium methoxide, use in
  carboxymethylation reactions, 147
Potassium persulfate
  use in free radical
    polymerization, 118,121
  use in methyl methacrylate
    polymerization, 121-123
Potassium phthalimide, reaction with
  perfluorobiphenyl, 130
Products, permanganate
  oxidations, 86-87
Propylene oxide,
  carboxyalkylation, 151-152
Pyridines, alkylation, 44

Q

Quaternary ammonium and phosphonium
  permanganates,
  solubilities, 83,84t
Quaternary ammonium ion,
  dealkylation, 158
Quaternary ammonium salts, as
  phase-transfer catalysts, 118-121
Quaternary onium salts
  as phase-transfer
    catalysts, 24-25,60
  polymer supported, formation and
    catalytic activity, 60

R

Racemic methylations, order in
  catalyst, 77-79

Rate
  methylation, initial versus delayed
    catalyst charge, 77,78f
  reaction of benzyl halides with
    potassium cyanide, 15-22
Rate constant, phase-transfer
  catalysis, 55
Rate law, methyl methacrylate
  polymerization, 122
Rate profiles, relation to 18-crown-6
  location in phase-transfer-
  catalyzed reactions, 19-22
Reaction mechanisms, permanganate
  oxidations, 86-93
Reaction profile, chiral phase
  transfer, 76f
Reactivity of anions, phase-transfer
  catalysis, 55-57
Reactivity scale, halides and
  pseudohalides, 56
Reduction, use of phase-transfer
  catalysts, 8-10
Resin-bound catalysts, activity, 4
Resonance stabilization, use to
  increase chemical resistance of
  ammonium salts, 43
Ring substitution, effect on catalytic
  activity, 61-62
Robinson annulation, chiral, 79,80f
Rubber particles, scission of
  polysulfide cross-links through
  phase-transfer catalysis, 155-167
Rubber peelings, devulcanization, 164

S

S_N2 reactions, activity of
  phase-transfer catalysts, 170-175
Sandwich ligands, catalytic activity
  in nucleophilic substitutions, 59t
Scission of polysulfide cross-links in
  rubber particles through
  phase-transfer catalysis, 155-167
Selectivity coefficients,
  phase-transfer catalysis, 57
Single-site phase-transfer catalysis,
  reaction of 4-chloronitrobenzene
  with 2,2,2-trifluoroethoxide
  ion, 176,178t
Sodium cation
  binding by nitrogen-pivot lariat
    ethers, 30,31f
  effect on polyvinylpyrrolidone-
    catalyzed reactions of
    cresoxides, 41t
  selectivity of crown ethers, 25-26
Sodium phenoxide, reaction with
  1,3-dibromopropane, 173,175t
Solid-liquid phase-transfer catalysis,
  aromatic substitution in
  condensation
  polymerization, 128-141

Solid-state complex structures, 32
Solubilities of permanganate
  salts, 83,84t
Soluble phase-transfer catalysts,
  reactivity and application, 54-65
Solvation, potassium permanganate by
  polyethers, 85t
Solvents
  effect of polarity on polymerization
    rate, 121-122
  effect on hexafluorobenzene
    polymerization, 133-134
  effect on perfluorobiphenyl
    polymerization, 133,135
  effect on permanganate
    oxidations, 92-93
Stability, ammonium salts, use of
  steric hindrance to increase, 43
Stability constants
  cations in methanol solutions
    containing crown ethers, 26t
  sodium ions in methanol
    solutions containing
    poly(ethylene glycols), 27
Stable catalysts for phase transfer at
  elevated temperatures, 38-51
Steric hindrance, stabilization of
  phase-transfer catalysts, 43
Styrene, polymerization, 121
Substituent effects, permanganate
  oxidations, 87-91
Synthesis of polymers by
  phase-transfer catalysis, 96-113

T

Taft plots, permanganate
  oxidations, 88-89f
Temperatures, elevated, stable
  catalysts for phase
  transfer, 38-51
Tetrabutylammonium bromide
  as a stable phase-transfer
    catalyst, 39-41
  use in ethyl acrylate
    polymerization, 120
  use in styrene polymerization, 121
Thermal transitions, polyethers and
  polyesters, 113f
Thermotropic polyethers, 107,112-113
Toluene, use in the reaction of benzyl
  halides with potassium
  cyanide, 15-22
Triblock copolymers
  gel permeation chromatograms, 111f
  H-1 NMR spectra, 111f
  synthesis, 109
Triphase catalysis
  general discussion, 60-61
  organometallic anion
    chemistry, 143-153

INDEX

W

Wacker process, description, 10
Water
  effect on bisphenol-A and
    hexafluorobenzene
    polymerization, 135-136,139f
  effect on 18-crown-6 concentration
    in toluene, 21t,22f
  effect on phase-transfer
    catalysis, 15-22,182
  effect on polymerization, 135-141

Water--Continued
  use in methyl methacrylate
    polymerization, 121
  use in permanganate oxidations, 91
Weak-nucleophile S_NAr reactions,
  activity of phase-transfer
  catalysts, 175-182

X

X-ray crystal structure analysis,
  lariat ethers, 32,33f

Production and indexing by Karen L. McCeney
Jacket design by Pamela Lewis

Elements typeset by Hot Type Ltd., Washington, DC
Printed and bound by Maple Press Co., York, PA

Recent ACS Books

Writing the Laboratory Notebook
By Howard M. Kanare
145 pages; clothbound ISBN 0-8412-0906-5

Polymeric Materials for Corrosion Control
Edited by Ray A. Dickie and F. Louis Floyd
ACS Symposium Series 322; 384 pp; ISBN 0-8412-0998-7

Porphyrins: Excited States and Dynamics
Edited by Martin Gouterman, Peter M. Rentzepis, and Karl D. Straub
ACS Symposium Series 321; 384 pp; ISBN 0-8412-0997-9

Agricultural Uses of Antibiotics
Edited by William A. Moats
ACS Symposium Series 320; 189 pp; ISBN 0-8412-0996-0

Fossil Fuels Utilization
Edited by Richard Markuszewski and Bernard D. Blaustein
ACS Symposium Series 319; 381 pp; ISBN 0-8412-0990-1

Materials Degradation Caused by Acid Rain
Edited by Robert Baboian
ACS Symposium Series 318; 449 pp; ISBN 0-8412-0988-X

Biogeneration of Aromas
Edited by Thomas H. Parliment and Rodney Croteau
ACS Symposium Series 317; 397 pp; ISBN 0-8412-0987-1

Formaldehyde Release from Wood Products
Edited by B. Meyer, B. A. Kottes Andrews, and R. M. Reinhardt
ACS Symposium Series 316; 240 pp; ISBN 0-8412-0982-0

Evaluation of Pesticides in Ground Water
Edited by Willa Y. Garner, Richard C. Honeycutt, and Herbert N. Nigg
ACS Symposium Series 315; 573 pp; ISBN 0-8412-0979-0

Water-Soluble Polymers: Beauty with Performance
Edited by J. E. Glass
Advances in Chemistry Series 213; 449 pp; ISBN 0-8412-0931-6

Historic Textile and Paper Materials: Conservation and Characterization
Edited by Howard L. Needles and S. Haig Zeronian
Advances in Chemistry Series 212; 464 pp; ISBN 0-8412-0900-6

For further information and a free catalog of ACS books, contact:
American Chemical Society, Sales Office
1155 16th Street, NW, Washington, DC 20036
Telephone 800-424-6747

ARY
Holl   642-3753